Springer Tracts in Advanced Robotics
Volume 27

Editors: Bruno Siciliano · Oussama Khatib · Frans Groen

Springer Tracts in Advanced Robotics

Edited by B. Siciliano, O. Khatib, and F. Groen

Michael Montemerlo · Sebastian Thrun

FastSLAM

A Scalable Method for the Simultaneous Localization and Mapping Problem in Robotics

With 50 Figures

Springer

Professor Bruno Siciliano, Dipartimento di Informatica e Sistemistica, Università degli Studi di Napoli Federico II, Via Claudio 21, 80125 Napoli, Italy, email: siciliano@unina.it

Professor Oussama Khatib, Robotics Laboratory, Department of Computer Science, Stanford University, Stanford, CA 94305-9010, USA, email: khatib@cs.stanford.edu

Professor Frans Groen, Department of Computer Science, Universiteit van Amsterdam, Kruislaan 403, 1098 SJ Amsterdam, The Netherlands, email: groen@science.uva.nl

Authors

Dr. Michael Montemerlo
Dr. Sebastian Thrun

Stanford University
Department of Computer Science
94305-9045 Stanford
USA

mmde@stanford.edu
thrun@stanford.edu

ISSN print edition: 1610-7438
ISSN electronic edition: 1610-742X

ISBN-10 3-540-46399-2 **Springer Berlin Heidelberg New York**
ISBN-13 978-3-540-46399-3 **Springer Berlin Heidelberg New York**

Library of Congress Control Number: 2006936725

Springer is a part of Springer Science+Business Media
springer.com

Typesetting: Digital data supplied by authors.
Data-conversion and production: PTP-Berlin Protago-TEX-Production GmbH, Germany (www.ptp-berlin.com)
Cover-Design: WMXDesign GmbH, Heidelberg
Printed on acid-free paper 89/3141/Yu - 5 4 3 2 1 0

STAR (Springer Tracts in Advanced Robotics) has been promoted under the auspices of EURON (European Robotics Research Network)

Foreword

At the dawn of the new millennium, robotics is undergoing a major transformation in scope and dimension. From a largely dominant industrial focus, robotics is rapidly expanding into the challenges of unstructured environments. Interacting with, assisting, serving, and exploring with humans, the emerging robots will increasingly touch people and their lives.

The goal of the new series of Springer Tracts in Advanced Robotics (STAR) is to bring, in a timely fashion, the latest advances and developments in robotics on the basis of their significance and quality. It is our hope that the wider dissemination of research developments will stimulate more exchanges and collaborations among the research community and contribute to further advancement of this rapidly growing field.

The monograph written by Michael Montemerlo and Sebastian Thrun is the second in the series devoted to one of the hottest research topics in the latest few years, namely Simultaneous Localization and Map Bulding (SLAM), which reached a peak of popularity on the occasion of the DARPA Grand Challenge in October 2005, remarkably won by the team headed by the authors.

The contents expand the first author's doctoral dissertation and describe a new family of fast algorithms based on the use of particle filters, which provide new insights into the data association problem that is paramount in SLAM. The algorithms have been successfully applied in different dynamic environments, including the solution to the problem of people tracking. No doubt, a very fine addition to the Series!

Naples, Italy
August 2006

Bruno Siciliano
STAR Editor

Preface

This monograph describes a new family of algorithms for the *simultaneous localization and mapping* problem in robotics (SLAM). SLAM addresses the problem of acquiring an environment map with a roving robot, while simultaneously localizing the robot relative to this map. This problem has received enormous attention in the robotics community in the past few years.

The FastSLAM family of algorithms applies particle filters to the SLAM problem. To accommodate the uncertainty in robot motion, the space of possible paths are approximated using sample paths. Each path obtains its own environment map. The resulting algorithm is efficient, and it provides new insights into the data association problem that is paramount in SLAM. FastSLAM-style algorithms have enabled robots to acquire maps of unprecedented size and accuracy, in a number of robot application domains.

This monograph is based on the first author's doctoral dissertation. The research would not have been possible without the close collaboration with Ben Wegbreit and Daphne Koller, who authored with us the original FastSLAM publications. It would also not be possible without the work of Kevin Murphy, who provided some essential mathematical insights upon which this work is built. A special thanks to Red Whittaker, Tony Stentz, Dieter Fox, Dirk Hähnel, Eduardo Nebot, and Wolfram Burgard, who contributed in many different ways to this research.

We are also indebted to DARPA and the Fannie and John Hertz Foundation for the financial support that made this work possible. Much of this work came together when we were both with Carnegie Mellon University, which we gratefully acknowledge.

Finally, we would like to thank our families – and especially Rosemary and Petra – for their love, support, and encouragement.

Stanford
July 2006

Michael Montemerlo
Sebastian Thrun

Contents

Table of Notations

s_t	pose of the robot at time t
s^t	complete path of the robot $\{s_1, s_2, \ldots, s_t\}$
θ_n	position of the n-th landmark
Θ	map, set of all n landmark positions
z_t	sensor observation at time t
z^t	set of all observations $\{z_1, z_2, \ldots, z_t\}$
u_t	robot control at time t
u^t	set of all controls $\{u_1 u_2, \ldots, u_t\}$
n_t	data association of observation at time t
n^t	set of all data associations $\{n_1, n_2, \ldots, n_t\}$
$h(s_{t-1}, u_t)$	vehicle motion model
P_t	linearized vehicle motion noise
$g(s_t, \Theta, n_t)$	vehicle measurement model
R_t	linearized vehicle measurement noise
$\hat{z}_{n_t}$	expected measurement of n_t-th landmark
$z_t - \hat{z}_{n_t}$	measurement innovation
Z_t	innovation covariance matrix
G_θ	Jacobian of measurement model with respect to landmark pose
G_{s_t}	Jacobian of measurement model with respect to robot pose
S_t	FastSLAM particle set at time t
$S_t^{[m]}$	m-th FastSLAM particle at time t
$\mu_{n,t}^{[m]}, \Sigma_{n,t}^{[m]}$	n-th landmark EKF (mean, covariance) in the m-th particle
$\mathcal{N}(x; \mu, \Sigma)$	Normal distribution over x with mean μ and covariance Σ
$w_t^{[m]}$	Importance weight of the m-th particle
N	Total number of landmarks
M	Total number of FastSLAM particles

1

Introduction

The problem of Simultaneous Localization and Mapping, or SLAM, has attracted immense attention in the robotics literature. SLAM addresses the problem of a mobile robot moving through an environment of which no map is available *a priori*. The robot makes relative observations of its ego-motion and of features in its environment, both corrupted by noise. The goal of SLAM is to reconstruct a map of the world and the path taken by the robot. SLAM is considered by many to be a key prerequisite to truly autonomous robots [85].

If the true map of the environment were available, estimating the path of the robot would be a straightforward localization problem [16]. Similarly, if the true path of the robot were known, building a map would be a relatively simple task [63, 86]. However, when both the path of the robot and the map are unknown, localization and mapping must be considered concurrently—hence the name *Simultaneous* Localization and Mapping.

1.1 Applications of SLAM

SLAM is an essential capability for mobile robots traveling in unknown environments where globally accurate position data (e.g. GPS) is not available. In particular, mobile robots have shown significant promise for remote exploration, going places that are too distant [34], too dangerous [88], or simply too costly to allow human access. If robots are to operate autonomously in extreme environments undersea, underground, and on the surfaces of other planets, they must be capable of building maps and navigating reliably according to these maps. Even in benign environments, such as the interiors of buildings, accurate, prior maps are often difficult to acquire. The capability to map an unknown environment allows a robot to be deployed with minimal infrastructure. This is especially important if the environment changes over time.

The maps produced by SLAM algorithms typically serve as the basis for motion planning and exploration. However, the maps often have value in their

(a) Undersea (b) Underground (c) Other planets

Fig. 1.1. Target environments for SLAM.

own right. In July of 2002, nine miners in the Quecreek Mine in Sommerset, Pennsylvania were trapped underground for three and a half days after accidentally drilling into a nearby abandoned mine. A subsequent investigation attributed the cause of the accident to inaccurate maps [32]. Since the accident, mobile robots and SLAM have been investigated as possible technologies for acquiring accurate maps of abandoned mines. One such robot, shown in Figure 1.1(b), is capable of building 3D reconstructions of the interior of abandoned mines using SLAM technology [88].

1.2 Joint Estimation

The chicken-or-egg relationship between localization and mapping is a consequence of how errors in the robot's sensor readings are corrupted by error in the robot's motion. As the robot moves, its pose estimate is corrupted by motion noise. The perceived locations of objects in the world are, in turn, corrupted by both measurement noise and the error in the estimated pose of the robot. Unlike measurement noise, however, error in the robot's pose will have a systematic effect on the error in the map. In general, this effect can be stated more plainly; *error in the robot's path correlates errors in the map.* As a result, the true map cannot be estimated without also estimating the true path of the robot. The relationship between localization and mapping was first identified by Smith and Cheeseman [82] in their seminal paper on SLAM in 1986.

Figure 1.2 shows a set of laser range scans collected by a mobile robot moving through a typical indoor environment. The robot generates estimates of its position using odometers attached to each of its wheels. In Figure 1.2(a), the laser scans are plotted with respect to the estimated position of the robot. Clearly, as error accumulates in the robot's odometry, the map becomes increasingly inaccurate. Figure 1.2(b) shows the laser readings plotted according to the path of the robot reconstructed by a SLAM algorithm.

Although the relationship between robot path error and map error does make the SLAM problem harder to solve in principle, one can exploit this

(a) Map built without correcting robot path

(b) Map built using SLAM

Fig. 1.2. Correlation between robot path error and map error.

relationship to factor the SLAM problem into a set of much smaller problems. Each of these smaller problems can be solved efficiently.

1.3 Posterior Estimation

Two types of information are available to the robot over time: controls and observations. Controls are noisy predictions of the robot's motion, while observations are noisy measurements of features in the robot's environment. Each control or observation, coupled with an appropriate noise model, can be thought of as a probabilistic constraint. Each control probabilistically constrains two successive poses of the robot. Observations, on the other hand, constrain the relative positions of the robot and objects in the map. When previously observed map features are revisited, the resulting constraints can be used to update not only the current map feature and robot pose, but also correct map features that were observed in the past. An example of a network of constraints imposed by controls and observations is shown in Figure 1.3.

Initially, the constraints imposed by controls and observations may be relatively weak. However, as map features are repeatedly observed, the constraints will become increasingly rigid. In the limit of an infinite number of observations and controls, the positions of all map features will become fully correlated [19]. The primary goal of SLAM is to estimate this true map and the true pose of the robot, given the currently available set of observations and controls.

One approach to the SLAM problem would be to estimate the most likely robot pose and map using a batch estimation algorithm similar to those used in the Structure From Motion literature [43, 91]. While extremely powerful, these techniques operate on the complete set of observations and controls, which grows without bound over time. As a consequence, these algorithms are not appropriate for online operation. Furthermore, these algorithms generally do not estimate the certainty with which different sections of the map are known, an important consideration for a robot exploring an unknown environment.

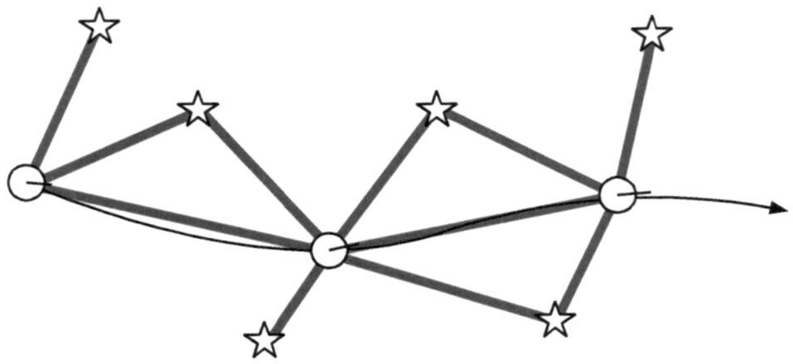

Fig. 1.3. Observations and controls form a network of probabilistic constraints on the pose of the robot and features in the robot's environment. These constraints are shown as thick lines.

The most popular online solutions to the SLAM problem attempt to estimate the posterior probability distribution over all possible maps Θ and robot poses s_t conditioned on the full set of controls u^t and observations z^t at time t. The observation at time t will be written as z_t, while the set of all observations up to time t will be written z^t. Similarly, the control at time t will be written u_t, and the set of all controls up to time t will be written u^t.

Using this notation, the joint posterior distribution over maps and robot poses can be written as:

$$p(s_t, \Theta | z^t, u^t) \tag{1.1}$$

This distribution is referred to as the SLAM posterior. At first glace, posterior estimation may seem even less feasible than the batch estimation approach. However, by making judicious assumptions about how the state of the world evolves, the SLAM posterior can be computed efficiently. Posterior estimation has several advantages over solutions that consider only the most likely state of the world. First, considering a distribution of possible solutions leads to more robust algorithms in noisy environments. Second, uncertainty can be used to compare the information conveyed by different components of the solution. One section of the map may be very uncertain, while other parts of the map are well known.

Any parameterized model can be chosen for the map Θ, however it is typically represented as a set of point features, or "landmarks" [19]. In a real implementation, landmarks may correspond to the locations of features extracted from sensors, such as cameras, sonars, or laser range-finder s. Throughout most of this book we assume the point landmark model, though other representations can be used. Higher order geometric features, such as line segments [70], have also been used to represent maps in SLAM.

The full SLAM posterior (1.1) assumes that the correspondences between the observations z_t and map features in Θ are unknown. If we make the restrictive assumption that the landmarks are uniquely identifiable, the SLAM posterior can be rewritten as:

$$p(s_t, \Theta \mid z^t, u^t, n^t) \tag{1.2}$$

where n_t is the identity of the landmark observed at time t, and n^t is the set of all landmark identities up to time t. The correspondences n^t are often referred to as "data associations." Initially in this book, we will describe how to solve the SLAM problem given known data association, a special case of the general SLAM problem. Later, we show how this restriction can be relaxed by maintaining multiple data association hypotheses.

The following recursive formula, known as the Bayes Filter for SLAM, can be used to compute the posterior (1.2) at time t, given the posterior at time $t-1$. A complete derivation of the Bayes Filter will be given in Chapter 2.

$$\begin{aligned} p(s_t, \Theta | z^t, u^t, n^t) &= \\ &\eta\, p(z_t | s_t, \Theta, n_t) \int p(s_t | s_{t-1}, u_t)\, p(s_{t-1}, \Theta | z^{t-1}, u^{t-1}, n^{t-1}) ds_{t-1} \end{aligned} \tag{1.3}$$

In general, the integral in (1.3) cannot be evaluated in closed form. However, this function can be computed by assuming a particular form for the posterior distribution. Many statistical estimation techniques, including the Kalman filter and the particle filter, are simply approximations of the general Bayes Filter.

1.4 The Extended Kalman Filter

The dominant approach to the SLAM problem was introduced in a seminal paper by Smith and Cheeseman [82] in 1986, and first developed into an implemented system by Moutarlier and Chatila [64, 65]. This approach uses the Extended Kalman Filter (EKF) to estimate the posterior over robot pose and maps. The EKF approximates the SLAM posterior as a high-dimensional Gaussian over all features in the map and the robot pose. The off-diagonal elements of the covariance matrix of this multivariate Gaussian encode the correlations between pairs of state variables. By estimating the covariance between all pairs of state variables, the EKF is expressive enough to represent the correlated errors that characterize the SLAM problem. An example of the EKF run on simulated data is shown in Figure 1.4(a). The corresponding covariance matrix, drawn as a correlation matrix, is shown in Figure 1.4(b). The darker the matrix element, the higher the correlation between the state variables corresponding to the element's row and column. While the EKF has become the dominant approach to SLAM, it suffers from two problems that complicate its application in large, real-world environments: quadratic complexity and sensitivity to failures in data association.

1.4.1 Quadratic Complexity

The first drawback of the EKF as a solution to the SLAM problem is computational complexity. Both the computation time and memory required by the

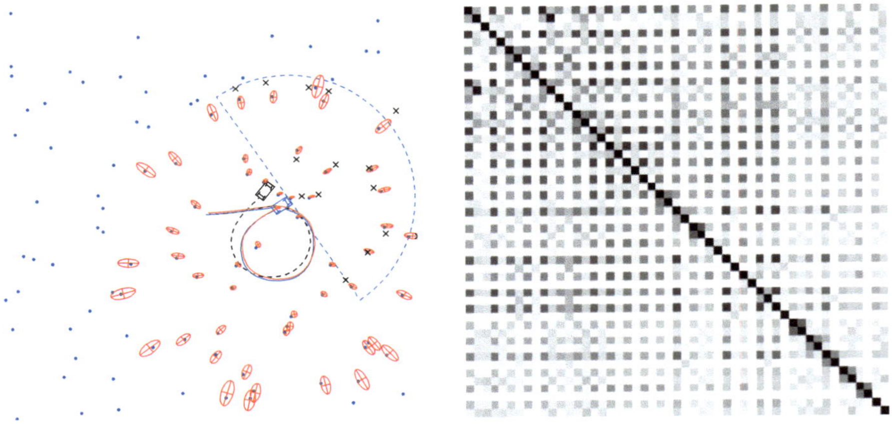

(a) SLAM in simulation. Ellipses represent the uncertainty in the positions of the landmarks.

(b) The corresponding covariance matrix (drawn as a correlation matrix). The darker the matrix element, the higher the correlation between the pair of variables.

Fig. 1.4. EKF applied to a simulated data set.

EKF scale quadratically with the number of landmarks in the map [70], limiting its application to relatively small maps. Quadratic complexity is a consequence of the Gaussian representation employed by the EKF. The uncertainty of the SLAM posterior is represented as a covariance matrix encoding the correlations between all possible pairs of state variables. In a two-dimensional world, the covariance matrix contains $2N + 3$ by $2N + 3$ entries, where N is the total number of landmarks in the map. Thus, it is easy to see how the memory required to store this covariance matrix grows with N^2.

Because the correlations between all pairs of state variables are maintained, any sensor observation incorporated into the EKF will necessarily affect all of the other state variables. To incorporate a sensor observation, the EKF algorithm must perform an operation on every element in the covariance matrix, which requires quadratic time.

In practice, the full EKF is rarely applied to the SLAM problem. The sensor update step can be made computationally tractable by using any of a variety of approximate EKF methods. These approximations will be discussed further in Section 1.5.

1.4.2 Single-Hypothesis Data Association

The second problem with EKF-based SLAM approaches is related to data association, the mapping between observations and landmarks. The SLAM problem is most commonly formulated given known data association, as in

(1.2). In the real world, the associations between observations and landmarks are hidden variables that must be determined in order to estimate the robot pose and the landmark positions.

The standard approach to data association in EKFs is to assign every observation to a landmark using a maximum likelihood rule; i.e. every observation is assigned to the landmark most likely to have generated it. If the probability of an observation belonging to an existing landmark is too low, it is considered for inclusion as a new landmark. Since the EKF has no mechanism for representing uncertainty over data associations, the effect of incorporating an observation given the wrong data association can never be undone. If a large number of readings are incorporated incorrectly into the EKF, the filter will diverge. Sensitivity to incorrect data association is a well known failure mode of the EKF [18].

The accuracy of data association in the EKF can be improved substantially by considering the associations of multiple observations simultaneously, at some computational cost [1, 68]. However, this does not address the underlying data association problem with the EKF; namely that it chooses a single data association hypothesis at every time step. The correct association for a given observation is not always the most probable choice when it is first considered. In fact, the true association for an observation may initially appear to be quite improbable. Future observations may be required to provide enough information to clearly identify the association as correct. Any EKF algorithm that maintains a single data association per time step, will inevitably pick wrong associations. If these associations can never be revised, repeated mistakes can cause the filter to diverge.

Multiple data association hypotheses can always be considered by maintaining multiple copies of the EKF, one for each probable data association hypothesis [77]. However, the computational and memory requirements of the EKF make this approach infeasible for the SLAM problem.

1.5 Structure and Sparsity in SLAM

At any given time, the observations and controls accumulated by the robot constrain only a small subset of the state variables. This sparsity in the dependencies between the data and the state variables can be exploited to compute the SLAM posterior in a more efficient manner. For example, two landmarks separated by a large distance are often weakly correlated. Moreover, nearby pairs of distantly separated landmarks will have very similar correlations. A number of approximate EKF SLAM algorithms exploit these properties by breaking the complete map into a set of smaller submaps. Thus, the large EKF can be decomposed into a number of loosely coupled, smaller EKFs. This approach has resulted in a number of efficient, approximate EKF algorithms that require linear time [36], or even constant time [1, 5, 7, 49] to incorporate sensor observations (given known data association).

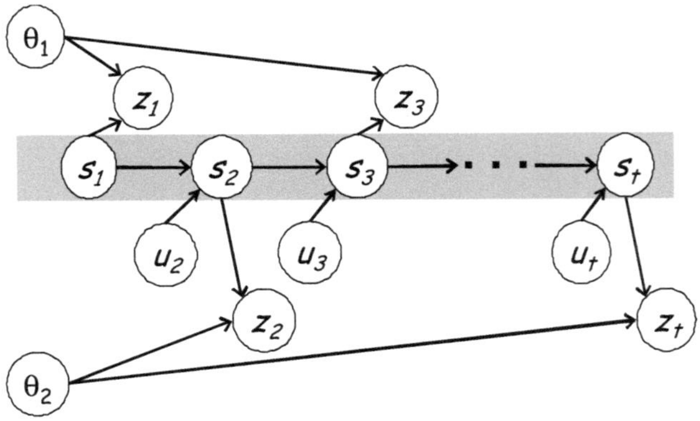

Fig. 1.5. SLAM as a Dynamic Bayes Network.

While *spatially* factoring the SLAM problem does lead to efficient EKF-based algorithms, the new algorithms face the same difficulties with data association as the original EKF algorithm. This book presents an alternative solution to the SLAM problem which exploits sparsity in the dependencies between state variables *over time*. In addition to enabling efficient computation of the SLAM posterior, this approach can maintain multiple data association hypotheses. The result is a SLAM algorithm that can be employed in large environments with significant data association ambiguity.

1.6 FastSLAM

As shown in Section 1.2, correlations between elements of the map *only* arise through robot pose uncertainty. Thus, if the robot's true path were known, the landmark positions could be estimated independently. Stated probabilistically, knowledge of the robot's true path renders estimates of landmark positions to be conditionally independent.

Proof of this statement can be seen by drawing the SLAM problem as a Dynamic Bayes Network, as shown in Figure 1.5. The robot's pose at time t is denoted s_t. This pose is a probabilistic function of the previous pose of the robot s_{t-1} and the control u_t executed by the robot. The observation at time t, written z_t, is likewise determined by the pose s_t and the landmark being observed θ_{n_t}. In the scenario depicted in Figure 1.5, the robot observes landmark 1 at $t = 1$ and $t = 3$, and observes landmark 2 at $t = 2$. The gray region highlights the complete path of the robot $s_1 \ldots s_t$. It is apparent from this network that this path "d-separates" [80] the nodes representing the two landmarks. In other words, if the true path of the robot is known, no information about the location of the first landmark can tell us anything about the location of the second landmark.

As a result of this relationship, the SLAM posterior given known data association (1.2) can be rewritten as the following product:

$$p(s^t, \Theta \mid z^t, u^t, n^t) = \underbrace{p(s^t \mid z^t, u^t, n^t)}_{\text{path posterior}} \underbrace{\prod_{n=1}^{N} p(\theta_n \mid s^t, z^t, u^t, n^t)}_{\text{landmark estimators}} \tag{1.4}$$

This factorization states that the full SLAM posterior can be decomposed into a product of $N + 1$ recursive estimators: one estimator over robot paths, and N independent estimators over landmark positions, each conditioned on the path estimate. This factorization was first presented by Murphy [66]. It is important to note that this factorization is exact, not approximate. It is a result of fundamental structure in the SLAM problem. A complete proof of this factorization will be given in Chapter 3.

The factored posterior (1.4) can be approximated efficiently using a particle filter [20, 51, 75], with each particle representing a sample path of the robot. Attached to each particle are N independent landmark estimators (implemented as EKFs), one for each landmark in the map. Since the landmark filters estimate the positions of individual landmarks, each filter is low dimensional. In total there are $N \cdot M$ Kalman filters. The resulting algorithm for updating this particle filter will be called FastSLAM. Readers familiar with the statistical literature should note that FastSLAM is an instance of the Rao-Blackwellized Particle Filter [23], by virtue of the fact that it combines a sampled representation with closed form calculations of certain marginals.

There are four steps to recursively updating the particle filter given a new control and observation, as shown in Table 1.1. The first step is to propose a new robot pose for each particle that is consistent with the previous pose and the new control. Next, the landmark filter in each particle that corresponds with the latest observation is updated using to the standard EKF update equations. Each particle is given an importance weight, and a new set of samples is drawn according to these weights. This importance resampling step corrects for the fact that the proposal distribution and the posterior distribution are not the same. This update procedure converges asymptotically to the true posterior distribution as the number of samples goes to infinity. In practice, FastSLAM generates a good reconstruction of the posterior with a relatively small number of particles (i.e. $M = 100$).

Initially, factoring the SLAM posterior using the robot's path may seem like a poor choice because the length of the path grows over time. Thus, one might expect the dimensionality of a filter estimating the posterior over robot paths to also grow over time. However, this is not the case for FastSLAM. As will be shown in Chapter 3, the landmark update equations and the importance weights only depend on the latest pose of the robot s_t, allowing us to silently forget the rest of the robot's path. As a result, each FastSLAM particle only needs to maintain an estimate of the current pose of the robot. Thus the dimensionality of the particle filter stays fixed over time.

Table 1.1. Basic FastSLAM Algorithm.

```
1. Sample a new robot path given the new control
2. Update landmark filters corresponding to the new observation
3. Assign a weight to each of the particles
4. Resample the particles according to their weights
```

1.6.1 Logarithmic Complexity

FastSLAM has two main advantages over the EKF. First, by factoring the estimation of the map into in separate landmark estimators conditioned on the robot path posterior, FastSLAM is able to compute the full SLAM posterior in an efficient manner. The motion update, the landmark updates, and the computation of the importance weights can all be accomplished in constant time per particle. The resampling step, if implemented naively, can be performed in linear time. However, this step can be implemented in logarithmic time by organizing each particle as a binary tree of landmark estimators, instead of an array. The $\log(N)$ FastSLAM algorithm can be used to build a map with over a million landmarks using a standard desktop computer.

1.6.2 Multi-hypothesis Data Association

Sampling over robot paths also has an important repercussion for determining the correct data associations. Since each FastSLAM particle represents a specific robot path, the same data association need not be applied to every particle. Data association decisions in FastSLAM can be made on a per-particle basis. Particles that predict the correct data association will tend to receive higher weights and be more likely to be resampled in the future. Particles that pick incorrect data associations will receive low weights and be removed. Sampling over data associations enables FastSLAM to revise past data associations as new evidence becomes available.

This same process also applies to the addition and removal of landmarks. Often, per-particle data association will lead to situations in which the particles build maps with differing numbers of landmarks. While this complicates the issue of computing the most probable map, it allows FastSLAM to remove spurious landmarks when more evidence is accumulated. If an observation leads to the creation a new landmark in a particular particle, but further observations suggest that the observation belonged to an existing landmark, then the particle will receive a low weight. This landmark will be removed from the filter when the improbable particle is not replicated in future resamplings. This process is similar in spirit to the "candidate lists" employed by EKF SLAM algorithms to test the stability of new landmarks [19, 50]. Unlike candidate lists, however, landmark testing in FastSLAM happens at no extra cost as a result of sampling over data associations.

1.7 Outline

This book presents an overview of the FastSLAM algorithm. Quantitative experiments will compare the performance of FastSLAM and the EKF on a variety of simulated and real world data sets. In Chapter 2, we formulate the SLAM problem and describe prior work in the field, concentrating primarily on EKF-based approaches. In Chapter 3, we describe the simplest version of the FastSLAM algorithm given both known and unknown data association. This version, which is called FastSLAM 1.0, is the simplest FastSLAM algorithm to implement and works well in typical SLAM environments. In Chapter 4, we present an improved version of the FastSLAM algorithm, called FastSLAM 2.0, that produces better results than the original algorithm. FastSLAM 2.0 incorporates the current observation into the proposal distribution of the particle filter and consequently produces more accurate results when motion noise is high relative to sensor noise. Chapter 4 also contains a proof of convergence for FastSLAM 2.0 in linear-Gaussian worlds. In Chapter 5, we describe a dynamic tracking problem that shares the same structure as the SLAM problem. We will show how a variation of the FastSLAM algorithm can be used to track dynamic objects from an imprecisely localized robot.

2

The SLAM Problem

In this chapter, we will present an overview of the Simultaneous Localization and Mapping problem, along with the most common SLAM approaches from the literature. Of primary interest will be the algorithms based on the Extended Kalman Filter.

2.1 Problem Definition

Consider a mobile robot moving through an unknown, static environment. The robot executes *controls* and collects *observations* of features in the world. Both the controls and the observations are corrupted by noise. Simultaneous Localization and Mapping (SLAM) is the process of recovering a map of the environment and the path of the robot from a set of noisy controls and observations.

If the path of the robot were known with certainty, then mapping would be a straightforward problem. The positions of objects in the robot's environment could be estimated using independent filters. However, when the path of the robot is unknown, error in the robot's path correlates errors in the map. As a result, the state of the robot and the map must be estimated *simultaneously*.

The correlation between robot pose error and map error can be seen graphically in Figure 2.1(a). A robot is moving along the path specified by the dashed line, observing nearby landmarks, drawn as circles. The shaded ellipses represent the uncertainty in the pose of the robot, drawn over time. As a result of control error, the robot's pose becomes more uncertain as the robot moves. The estimates of the landmark positions are shown as unshaded ellipses. Clearly, as the robot's pose becomes more uncertain, the uncertainty in the estimated positions of newly observed landmarks also increases.

In Figure 2.1(b), the robot completes the loop and revisits a previously observed landmark. Since the position of this first landmark is known with high accuracy, the uncertainty in the robot's pose estimate will decrease significantly. This newly discovered information about the robot's pose increases

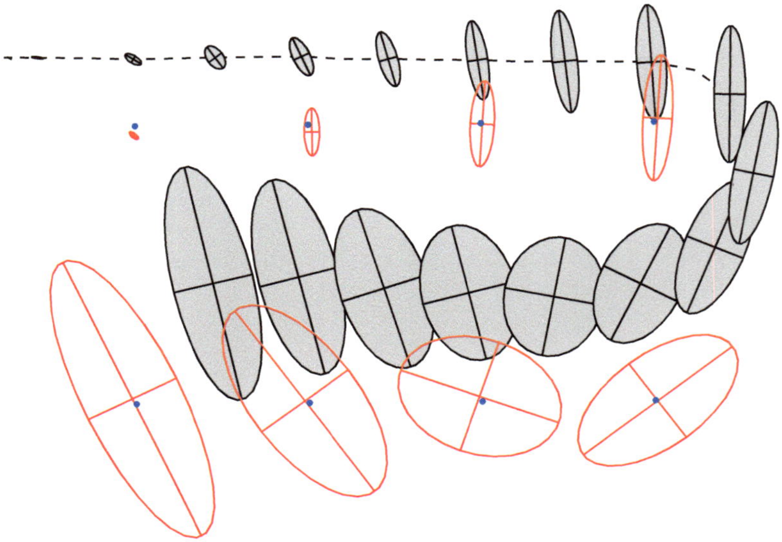

(a) Before closing the loop: landmark uncertainty increases as robot pose uncertainty increases. Robot pose estimates over time are shown as shaded ellipses. Landmark estimates are shown as unshaded ellipses.

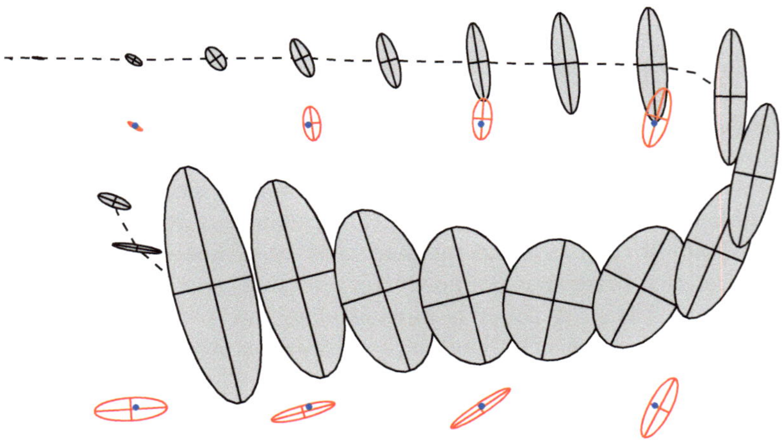

(b) After closing the loop: revisiting a known landmark decreases not only the robot pose uncertainty, but also the uncertainty of landmarks previously observed.

Fig. 2.1. Robot motion error correlates errors in the maps.

the certainty with which past poses of the robot are known as well. This, in turn, reduces the uncertainty of landmarks previously observed by the robot. The reader may notice that the shaded ellipses before the loop closure in Figure 2.1(b) do not shrink because they depict a time series of the robot's pose uncertainty and not revised estimates of the robot's past poses.

The effect of the observation on all of the landmarks around the loop is a consequence of the correlated nature of the SLAM problem. Errors in the map are correlated through errors in the robot's path. Any observation that provides information about the pose of the robot, will necessarily provide information about all previously observed landmarks.

2.2 SLAM Posterior

The pose of the robot at time t will be denoted s_t. For a robot operating in a planar environment, this pose consists of the robot's x-y position in the plane and its heading direction. All experimental results presented in this book were generated in planar environments, however the algorithms apply equally well to three-dimensional worlds. The complete trajectory of the robot, consisting of the robot's pose at every time step, will be written as s^t.

$$s^t = \{s_1, s_2, \ldots, s_t\} \tag{2.1}$$

We shall further assume that the robot's environment can be modeled as a set of N immobile, point landmarks. Point landmarks are commonly used to represent the locations of features extracted from sensor data, such as geometric features in a laser scan or distinctive visual features in a camera image. The set of N landmark locations will be written $\{\theta_1, \ldots, \theta_N\}$. For notational simplicity, the entire map will be written as Θ.

As the robot moves through the environment, it collects relative information about its own motion. This information can be generated using odometers attached to the wheels of the robot, inertial navigation units, or simply by observing the control commands executed by the robot. Regardless of origin, any measurement of the robot's motion will be referred to generically as a control. The control at time t will be written u_t. The set of all controls executed by the robot will be written u^t.

$$u^t = \{u_1, u_2, \ldots, u_t\} \tag{2.2}$$

As the robot moves through its environment, it observes nearby landmarks. In the most common formulation of the planar SLAM problem, the robot observes both the range and bearing to nearby obstacles. The observation at time t will be written z_t. The set of all observations collected by the robot will be written z^t.

$$z^t = \{z_1, z_2, \ldots, z_t\} \tag{2.3}$$

It is commonly assumed in the SLAM literature that sensor measurements can be decomposed into information about individual landmarks, such that each landmark observation can be incorporated independently from the other measurements. This is a realistic assumption in virtually all successful SLAM implementations, where landmark features are extracted one-by-one from raw sensor data. Thus, we will assume that each observation provides information about the location of exactly one landmark θ_n relative to the robot's current pose s_t. The variable n represents the identity of the landmark being observed. In practice, the identities of landmarks usually can not be observed, as many landmarks may look alike. The identity of the landmark corresponding to the observation z_t will be written as n_t, where $n_t \in \{1, \ldots, N\}$. For example, $n_8 = 3$ means that at time $t = 8$ the robot observed the third landmark. Landmark identities are commonly referred to as "data associations" or "correspondences." The set of all data associations will be written n^t.

$$n^t = \{n_1, n_2, \ldots, n_t\} \tag{2.4}$$

Again for simplicity, we will assume that the robot receives exactly one measurement z_t and executes exactly one control u_t per time step. Multiple observations per time step can be processed sequentially, but this leads to a more cumbersome notation.

Using the notation defined above, the primary goal of SLAM is to recover the best estimate of the robot pose s_t and the map Θ, given the set of noisy observations z^t and controls u^t. In probabilistic terms, this is expressed by the following posterior:

$$p(s_t, \Theta \mid z^t, u^t) \tag{2.5}$$

If the set of data associations n^t is also given, the posterior can be rewritten as:

$$p(s_t, \Theta \mid z^t, u^t, n^t) \tag{2.6}$$

2.3 SLAM as a Markov Chain

The SLAM problem can be described best as a probabilistic Markov chain. A graphical depiction of this Markov chain is shown in Figure 2.2. The current pose of the robot s_t can be written as a probabilistic function of the pose at the previous time step s_{t-1} and the control u_t executed by the robot. This function is referred to as the *motion model* because it describes how controls drive the motion of the robot. Additionally, the motion model describes how noise in the controls injects uncertainty into the robot's pose estimate. The motion model is written as:

$$p(s_t \mid s_{t-1}, u_t) \tag{2.7}$$

Sensor observations gathered by the robot are also governed by a probabilistic function, commonly referred to as the *measurement model*. The observation z_t

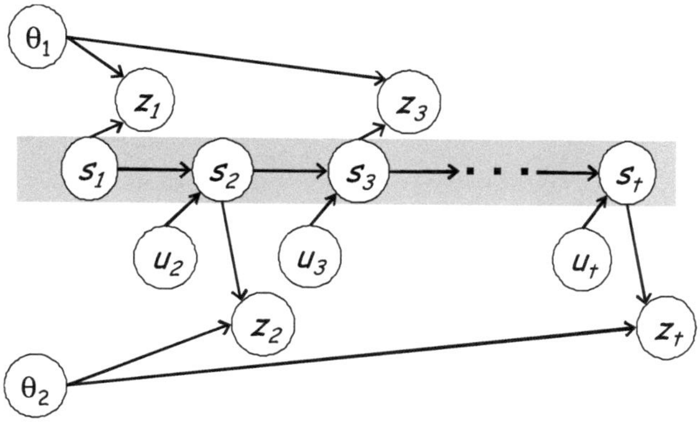

Fig. 2.2. SLAM as a Dynamic Bayes Network.

is a function of the observed landmark θ_{n_t} and the pose of the robot s_t. The measurement model describes the physics and the error model of the robot's sensor. The measurement model is written as:

$$p(z_t \mid s_t, \Theta, n_t) \tag{2.8}$$

Using the motion model and the measurement model, the SLAM posterior at time t can be computed recursively as function of the posterior at time $t-1$. This recursive update rule, known as the Bayes filter for SLAM, is the basis for the majority of online SLAM algorithms.

2.3.1 Bayes Filter Derivation

The Bayes Filter can be derived from the SLAM posterior as follows. First, the posterior (2.6) is rewritten using Bayes Rule.

$$p(s_t, \Theta \mid z^t, u^t, n^t) = \eta\, p(z_t \mid s_t, \Theta, z^{t-1}, u^t, n^t)\, p(s_t, \Theta \mid z^{t-1}, u^t, n^t) \tag{2.9}$$

The denominator from Bayes rule is a normalizing constant and is written as η. Next, we exploit the fact that z_t is solely a function of the pose of the robot s_t, the map Θ, and the latest data association n_t, previously described as the measurement model. Hence the posterior becomes:

$$= \eta\, p(z_t \mid s_t, \Theta, n_t)\, p(s_t, \Theta \mid z^{t-1}, u^t, n^t) \tag{2.10}$$

Now we use the Theorem of Total Probability to condition the rightmost term of (2.10) on the pose of the robot at time $t-1$.

$$= \eta\, p(z_t \mid s_t, \Theta, n_t) \int p(s_t, \Theta \mid s_{t-1}, z^{t-1}, u^t, n^t)\, p(s_{t-1} \mid z^{t-1}, u^t, n^t) ds_{t-1} \tag{2.11}$$

The leftmost term inside the integral can be expanded using the definition of conditional probability.

$$= \eta\, p(z_t \mid s_t, \Theta, n_t) \tag{2.12}$$
$$\int p(s_t \mid \Theta, s_{t-1}, z^{t-1}, u^t, n^t)\, p(\Theta \mid s_{t-1}, z^{t-1}, u^t, n^t)\, p(s_{t-1} \mid z^{t-1}, u^t, n^t) ds_{t-1}$$

The first term inside the integral can now be simplified by noting that s_t is only a function of s_{t-1} and u_t, previously described as the motion model.

$$= \eta\, p(z_t \mid s_t, \Theta, n_t) \tag{2.13}$$
$$\int p(s_t \mid s_{t-1}, u_t)\, p(\Theta \mid s_{t-1}, z^{t-1}, u^t, n^t)\, p(s_{t-1} \mid z^{t-1}, u^t, n^t) ds_{t-1}$$

At this point, the two rightmost terms in the integral can be combined.

$$= \eta\, p(z_t \mid s_t, \Theta, n_t) \int p(s_t \mid s_{t-1}, u_t) p(s_{t-1}, \Theta \mid z^{t-1}, u^t, n^t) ds_{t-1} \tag{2.14}$$

Since the current pose u_t and data association n_t provide no new information about s_{t-1} or Θ without the latest observation z_t, they can be dropped from the rightmost term of the integral. The result is a recursive formula for computing the SLAM posterior at time t given the SLAM posterior at time $t-1$, the motion model $p(s_t \mid s_{t-1}, u_t)$, and the measurement model $p(z_t \mid s_t, \Theta, n_t)$.

$$p(s_t, \Theta \mid z^t, u^t, n^t) = \tag{2.15}$$
$$\eta\, p(z_t \mid s_t, \Theta, n_t) \int p(s_t \mid s_{t-1}, u_t)\, p(s_{t-1}, \Theta \mid z^{t-1}, u^{t-1}, n^{t-1}) ds_{t-1}$$

2.4 Extended Kalman Filtering

In general, the integral in the recursive update equation (2.15) cannot be computed in closed form. However, approximate SLAM algorithms have been developed by restricting the form of the SLAM posterior, the motion model, and the measurement model. Many of the original SLAM algorithms originate from a seminal paper by Smith and Cheesman [82], which proposed the use of the Extended Kalman Filter (EKF) to estimate the SLAM posterior.

The EKF represents the SLAM posterior as a high-dimensional, multivariate Gaussian parameterized by a mean μ_t and a covariance matrix Σ_t. The mean describes the most likely state of the robot and landmarks, and the covariance matrix encodes the pairwise correlations between all pairs of state variables.

$$p(s_t, \Theta \mid u^t, z^t, n^t) = \mathcal{N}(x_t; \mu_t, \Sigma_t) \tag{2.16}$$
$$x_t = \{s_t, \theta_1, \ldots, \theta_N\} \tag{2.17}$$

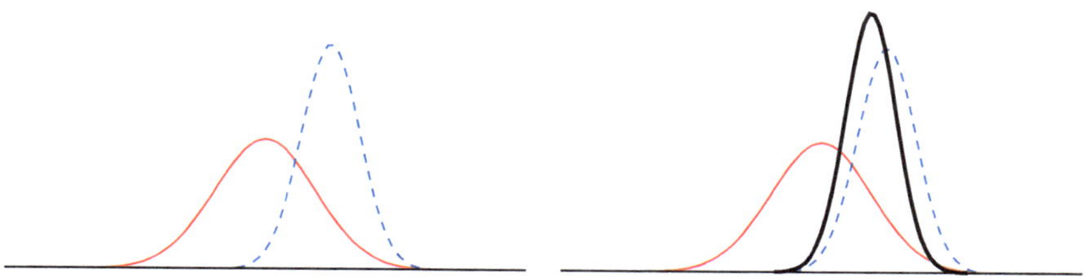

(a) Prior belief (solid line) and a new observation (dashed line)

(b) Belief after incorporating the new observation (thick line)

Fig. 2.3. One-Dimensional Kalman Filter.

$$\mu_t = \{\mu_{s_t}, \mu_{\theta_{1,t}}, \ldots, \mu_{\theta_{N,t}}\} \tag{2.18}$$

$$\Sigma_t = \begin{bmatrix} \Sigma_{s_t,t} & \Sigma_{s_t\theta_1,t} & \ldots & \Sigma_{s_t\theta_N,t} \\ \Sigma_{\theta_1 s_t,t} & \Sigma_{\theta_1,t} & \Sigma_{\theta_1\theta_2,t} & \\ \vdots & \Sigma_{\theta_2\theta_1,t} & \ddots & \\ \Sigma_{\theta_N s_t,t} & & & \Sigma_{\theta_N,t} \end{bmatrix} \tag{2.19}$$

For robots that move in a plane, the mean vector μ_t is of dimension $2N+3$, where N is the number of landmarks. Three dimensions are required to represent the pose of the robot, and two dimensions are required to specify the position of each landmark. Likewise, the covariance matrix is of size $2N+3$ by $2N+3$. Thus, the number of parameters needed to describe the EKF posterior is quadratic in the number of landmarks in the map.

Figure 2.3 shows a simple example of a Kalman Filter estimating the position of a single landmark in one dimension. Figure 2.3(a) shows the current belief in the landmark position (the solid distribution) and a new, noisy observation of the landmark (the dashed distribution). The Kalman Filter describes the optimal procedure for combining Gaussian beliefs in linear systems. In this case, the new posterior after incorporating the dashed observation is shown as a thick line in Figure 2.3(b).

The basic Kalman Filter algorithm is the optimal estimator for a linear system with Gaussian noise [3]. As its name suggests, the EKF is simply an extension of the basic Kalman Filter algorithm to non-linear systems. The EKF does this by replacing the motion and measurement models with non-linear models that are "linearized" around the most-likely state of the system. In general, this approximation is good if the true models are approximately linear and if the discrete time step of the filter is small.

The motion model will be written as the non-linear function $h(x_{t-1}, u_t)$ with linearized noise covariance P_t. Similarly, the measurement model will be written as the non-linear function $g(x_t, n_t)$ with linearized noise covariance

R_t. The EKF update equations can be written as follows:

$$\mu_t^- = h(\mu_{t-1}, u_t) \tag{2.20}$$
$$\Sigma_t^- = \Sigma_{t-1} + P_t \tag{2.21}$$
$$G_x = \nabla_{x_t} g(x_t, n_t)|_{x_t=\mu_t^-; n_t=n_t} \tag{2.22}$$
$$Z_t = G_x \Sigma_t^- G_x^T + R_t \qquad \hat{z}_{n_t} = g(\mu_t^-, n_t) \tag{2.23}$$
$$K_t = \Sigma_t^- G_x^T Z_t^{-1} \tag{2.24}$$
$$\mu_t = \mu_t^- + K_t(z_t - \hat{z}_{n_t}) \tag{2.25}$$
$$\Sigma_t = (I - K_t G_t)\Sigma_t^- \tag{2.26}$$

For a complete derivation of the Kalman Filter, see [46, 31, 87], and for a gentle introduction to the use of the Kalman Filter and the EKF, see [95]. It is important to note that if the SLAM problem is linear and Gaussian, then the Kalman Filter is both guaranteed to converge [69] and provably optimal [3]. Real-world SLAM problems are rarely linear, yet the EKF still tends to produce very good results in general. For this reason, the EKF is often held up as the "gold standard" of comparison for online SLAM algorithms.

The EKF has two substantial disadvantages when applied to the SLAM problem: quadratic complexity and sensitivity to failures in data association. The number of mathematical operations required to incorporate a control and an observation into the filter is dominated by the final EKF update equation (2.26). In the planar case, both K_t and G_t^T are of dimension $2N + 3$ by the dimensionality of the observation (typically two). Thus, the inner product in the calculation of Σ_t requires a number of calculations quadratic with the number of landmarks N.

The second problem with the EKF applies in situations in which the data associations n_t are unknown. The EKF maintains a single data association hypothesis per observation, typically chosen using a maximum likelihood heuristic. If the probability of an observation coming from any of the current landmarks is too low, the possibility of a new landmark is considered. If the data association chosen by this heuristic is incorrect, the effect of incorporating this observation into the EKF can never be removed. If many observations are incorporated into the EKF with wrong data associations, the EKF will diverge. This is a well known failure mode of the EKF [18]. The following sections will describe alternative approaches to SLAM that address the issues of efficient scaling and robust data association.

2.5 Scaling SLAM Algorithms

2.5.1 Submap Methods

While the Kalman Filter is the optimal solution to the linear-Gaussian SLAM problem, it is computationally infeasible for large maps. As a result, a great deal of SLAM research has concentrated on developing SLAM algorithms that approximate the performance of the EKF, but scale to much larger environments. The computational complexity of the EKF stems from the fact that covariance matrix Σ_t represents every pairwise correlation between the state variables. Incorporating an observation of a single landmark will necessarily have an affect on every other state variable.

Typically, the observation of a single landmark will have a weak effect on the positions of distant landmarks. For this reason, many researchers have developed EKF-based SLAM algorithms that decompose the global map into smaller submaps. One set of approaches exploits the fact that the robot may linger for some period of time in a small section of the global map. Postponement [10, 47], and the Compressed Extended Kalman Filter (CEKF) [36], are both techniques that delay the incorporation of local information into the global map while the robot stays inside a single submap. These techniques are still optimal, in that they generate the same results as the full EKF. However, the computation required by the two algorithms is reduced by a constant factor because the full map updates are performed less frequently.

Breaking the global map into submaps can also lead to a more sparse description of the correlations between map elements. Increased sparsity can be exploited to compute more efficient sensor updates. Network Coupled Feature Maps [1], ATLAS [5], the Local Mapping Algorithm [7], and the Decoupled Stochastic Mapping [49] frameworks all consider relationships between a sparse network of submaps. When the robot moves out of one submap, it either creates a new submap or relocates itself in a previously defined submap. Each approach reduces the computational requirement of incorporating an observation to constant time, given known data association. However, these computational gains come at the cost of slowing down the overall rate of convergence. Each map has far fewer features than the overall map would have, and the effects of observations on distant landmarks may have to percolate through multiple correlation links.

Guivant and Nebot presented a similar method called Suboptimal SLAM [36], in which the local maps are all computed with respect to a small number of base landmarks. Since the different constellations of landmarks are kept in different coordinate frames, they can be decorrelated more easily than if every landmark were in a single coordinate frame. The resulting algorithm produces an estimate that is an approximation to the true EKF estimate, however it requires linear time and memory.

2.5.2 Sparse Extended Information Filters

Another filter approach to decomposing the SLAM problem is to represent maps using potential functions between nearby landmarks, similar to Markov Random Fields [6]. One such approach is the Sparse Extended Information Filter (SEIF) proposed in [90]. SEIFs implement an alternate parameterization of the Kalman Filter, called the Information Filter. Instead of updating a covariance matrix Σ, SEIFs update Σ^{-1}, the precision matrix. This parameterization is useful because the precision matrix is sparse if correlations are maintained only between nearby landmarks. Under appropriate approximations, this technique has been shown to provide efficient updates (given known data association) with a linear memory requirement.

2.5.3 Thin Junction Trees

The Thin Junction Tree Filter (TJTF) of Paskin [73] is a SLAM algorithm based on the same principle as the SEIF. Namely, maintaining a sparse network of probabilistic constraints between state variables enables efficient inference. The TJTF represents the SLAM posterior using a graphical model called a junction tree. The size of the junction tree grows as new landmarks are added to the map, but it can be "thinned" using an operation called variable contraction. The thinning operation can be viewed as a method for making the precision matrix of a SEIF sparse, however global maps can be extracted from TJTFs without any matrix inversion. TJTFs require linear computation in general, which can be reduced to constant time with further approximation.

2.5.4 Covariance Intersection

SLAM algorithms that treat correlated variables as if they were independent will necessarily underestimate their covariance. Underestimated covariance can lead to divergence and make data association extremely difficult. Ulmann and Juiler present an alternative to maintaining the complete joint covariance matrix called Covariance Intersection [45]. Covariance Intersection updates the landmark position variances conservatively, in such a way that allows for all possible correlations between the observation and the landmark. Since the correlations between landmarks no longer need to be maintained, the resulting SLAM algorithm requires linear time and memory. Unfortunately, the landmark estimates tend to be extremely conservative, leading to extremely slow convergence and highly ambiguous data association.

2.5.5 Graphical Optimization Methods

There exists a family of off-line SLAM algorithms which treat the SLAM problems as optimization problem. These techniques exploit a similar independence as discussed in the previous chapter. By retaining all past poses, the set of constraints between different variables in the SLAM posterior can be represented by a set of sparse links. Relaxing those links using conventional optimization techniques then leads to a most likely map and a most likely robot path; assuming known correspondence. Possibly the earliest work on this paradigm is by Lu and Milios [53], which was first implemented by Gutmann [38]. Golfarelli et al. [33] established the relation of SLAM problems and spring-mass models, and Duckett et al. [24] provided a first efficient technique for solving such problems; see also subsequent work in [48, 58]. The relation between covariances and the information matrix is discussed in [29] and [15]. Folkesson and Christensen introduced a term "Graphical SLAM" into the literature [28], for a related graphical relaxation method. A state-of-the-art discussion of graphical optimization techniques can be found in [87]. The optimization-based graphical methods are usually offline, and as such are related to the Structure From Motion literature [43, 91].

2.6 Robust Data Association

In real SLAM applications, the data associations n^t are rarely observable. However, if the uncertainty in landmark positions is low relative to the average distance between landmarks, simple heuristics for determining the correct data association can be quite effective. In particular, the most common approach to data association in SLAM is to assign each observation using a maximum likelihood rule. In other words, each observation is assigned to the landmark most likely to have generated it. If the maximum probability is below some fixed threshold, the observation is considered for addition as a new landmark.

In the case of the EKF, the probability of the observation can be written as a function of the difference between the observation z_t and the expected observation $\hat{z}_{n_t}$. This difference is known as the "innovation."

$$\hat{n}_t = \underset{n_t}{\operatorname{argmax}}\, p(z_t \mid n_t, s^t, z^{t-1}, u^t, \hat{n}^{t-1}) \tag{2.27}$$

$$= \underset{n_t}{\operatorname{argmax}} \frac{1}{\sqrt{|2\pi Z_t|}} \exp\left\{-\frac{1}{2}(z_t - \hat{z}_{n_t})^T Z_t^{-1}(z_t - \hat{z}_{n_t})\right\} \tag{2.28}$$

This data association heuristic is often reformulated in terms of negative log likelihood, as follows:

$$\hat{n}_t = \underset{n_t}{\operatorname{argmin}} \ln|Z_t| + (z_t - \hat{z}_{n_t})^T Z_t^{-1}(z_t - \hat{z}_{n_t}) \tag{2.29}$$

The second term of this equation is known as Mahalanobis distance [80], a distance metric normalized by the covariances of the observation and the landmark estimate. For this reason, data association using this metric is often referred to as "nearest neighbor" data association [3], or nearest neighbor gating.

Maximum likelihood data association generally works well when the correct data association is significantly more probable than the incorrect associations. However, if the uncertainty in the landmark positions is high, more than one data association will receive high probability. If a wrong data association is picked, this decision can have a catastrophic result on the accuracy of the resulting map. This kind of data association ambiguity can be induced easily if the robot's sensors are very noisy.

One approach to this problem is to only incorporate observations that lead to unambiguous data associations (i.e. if only one data association falls within the nearest neighbor threshold). However, if the SLAM environment is noisy, a large percentage of the observations will go unprocessed. Moreover, failing to incorporate observations will lead to overestimated landmark covariances, which makes future data associations even more ambiguous.

A number of more sophisticated approaches to data association have been developed in order to deal with ambiguity in noisy environments.

2.6.1 Local Map Sequencing

Tardos et al. [84] developed a technique called Local Map Sequencing for building maps of indoor environments using sonar data. Sonar sensors tend to be extremely noisy and viewpoint dependent. The Local Map Sequencing algorithm collects a large number of sonar readings as the robot moves over a short distance. These readings are processed by two Hough transforms [2] that detect corners and line segments in the robot's vicinity given the entire set of observations. Features from the Hough transform are used to build a map of the robot's local environment. Multiple local maps are then pieced together to build a global map of the world.

The Hough transforms make the data association robust because multiple sensor readings, taken from different robot poses, vote to determine the correct interpretation of the data. Using this approach, reasonably accurate maps can be built with inexpensive, noisy sensors. The authors also suggest RANSAC [27] as another voting algorithm to determine data association with noisy sensors.

2.6.2 Joint Compatibility Branch and Bound

If multiple observations are gathered per control, the maximum likelihood approach will treat each data association decision as a independent problem. However, because data association ambiguity is caused in part by robot pose uncertainty, the data associations of simultaneous observations are correlated. Considering the data association of each of the observations separately also ignores the issue of mutual exclusion. Multiple observations cannot be associated with the same landmark during a single time step.

Neira and Tardos [68] showed that both of these problems can be remedied by considering the data associations of all of the observations simultaneously, much like the Local Map Sequencing algorithm does. Their algorithm, called Joint Compatibility Branch and Bound (JCBB), traverses the Interpretation Tree [35], which is the tree of all possible joint correspondences. Different joint data association hypotheses are compared using joint compatibility, a measure of the probability of the set of observations occurring together. In the EKF framework, this can be computed by finding the probability of the joint innovations of the observations. Clearly, considering joint correspondences comes at some computational cost, because an exponential number of different hypotheses must be considered. However, Neira and Tardos showed that many of these hypotheses can be excluded without traversing the entire tree.

2.6.3 Combined Constraint Data Association

Bailey [1] presented a data association algorithm similar to JCBB called Combined Constraint Data Association (CCDA). Instead of building a tree of joint correspondences, CCDA constructs a undirected graph of data association constraints, called a "Correspondence Graph". Each node in the graph, represents a candidate pairing of observed features and landmarks, possibly determined using a nearest neighbor test. Edges between the nodes represent joint compatibility between pairs of data associations. The algorithm picks the set of joint data associations that correspond to the largest clique in the correspondence graph. The results of JCBB and CCDA should be similar, however the CCDA algorithm is able to determine viable data associations when the pose of the robot relative to the map is completely unknown.

2.6.4 Iterative Closest Point

Scan matching [54] is a data association method that based on a modified version of the Iterative Closest Point (ICP) algorithm [4]. This algorithm alternates between a step in which correspondences between data are identified, and a step in which a new robot path is recovered from the current correspondences. This iterative optimization is similar in spirit to Expectation Maximization (EM) [17] and RANSAC [27]. First, a locally consistent map is built using scan-matching [39], a maximum likelihood mapping approach. Next, observations are matched between different sensor scans using a distance metric. Based on the putative correspondences, a new set of robot poses is derived. This alternating process is iterated several times until some convergence criterion is reached. This process has shown significant promise for the data association problems encountered in environments with very large loops.

2.6.5 Multiple Hypothesis Tracking

Thus far, all of the data association algorithms presented all choose a single data association hypothesis to be fed into an EKF, or approximate EKF algorithm. There are a few algorithms that maintain multiple data association hypotheses over time. This is especially useful if the correct data association of an observation cannot be inferred from a single measurement. One such approach in the target tracking literature is the Multiple Hypothesis Tracking or MHT algorithm [77]. MHT maintains a set of hypothesized tracks of multiple targets. If a particular observation has multiple, valid data association interpretation s, new hypotheses are created according to each hypothesis. In order to keep the number of hypotheses from expanding without bound, heuristics are used to prune improbable hypotheses from the set over time.

Maintaining multiple EKF hypotheses for SLAM is unwieldy because each EKF maintains a belief over robot pose and the entire map. Nebot et al. [67] have developed a similar technique that "pauses" map-building when data association becomes ambiguous, and performs multi-hypothesis localization using a particle filter until the ambiguity is resolved. Since map building is not performed when there is data association ambiguity, the multiple hypotheses are over robot pose, which is a low-dimensional quantity. However, this approach only works if data association ambiguity occurs sporadically. This can be useful for resolving occasional data association problems when closing loops, however the algorithm will never spend any time mapping if the ambiguity is persistent.

2.7 Comparison of FastSLAM to Existing Techniques

The remainder of this book will describe FastSLAM, an alternative approach to estimating the SLAM posterior. Unlike submap EKF approaches, which factor the SLAM problem spatially, FastSLAM factors the SLAM posterior over time using the path of the robot. The resulting algorithm scales logarithmically with the number of landmarks in the map, which is sufficient to process maps with millions of features.

FastSLAM samples over potential robot paths, instead of maintaining a parameterized distribution of solutions like the EKF. This enables FastSLAM to apply different data association hypotheses to different solutions represented under the SLAM posterior. FastSLAM, therefore, maintains the multiple-hypothesis tracking abilities of MHT and the hybrid filter approach, yet it can perform localization and mapping simultaneously, even with consistently high data association ambiguity.

3

FastSLAM 1.0

In this chapter we will describe the basic FastSLAM algorithm, an alternative approach to SLAM that is based on particle filtering.

The FastSLAM algorithm is based on a structural property of the SLAM problem that the EKF fails to exploit. Each control or observation collected by the robot only constrains a small number of state variables. Controls probabilistically constrain the pose of the robot relative to its previous pose, while observations constrain the positions of landmarks relative to the robot. It is only after a large number of these probabilistic constraints are incorporated that the map becomes fully correlated. The EKF, which makes no assumptions about structure in the state variables, fails to take advantage of this sparsity over time.

FastSLAM exploits conditional independences that are a consequence of the sparse structure of the SLAM problem to factor the posterior into a product of low dimensional estimation problems. The resulting algorithm scales efficiently to large maps and is robust to significant ambiguity in data association.

3.1 Particle Filtering

The Kalman Filter and the EKF represent probability distributions using a parameterized model (a multivariate Gaussian). Particle filters, on the other hand, represent distributions using a finite set of sample states, or "particles" [20, 51, 75]. Regions of high probability contain a high density of particles, whereas regions of low probability contain few or no particles. Given enough samples, this non-parametric representation can approximate arbitrarily complex, multi-modal distributions. In the limit of an infinite number of samples, the true distribution can be reconstructed exactly [21], under some very mild assumptions. Given this representation, the Bayes Filter update equation can be implemented using a simple sampling procedure.

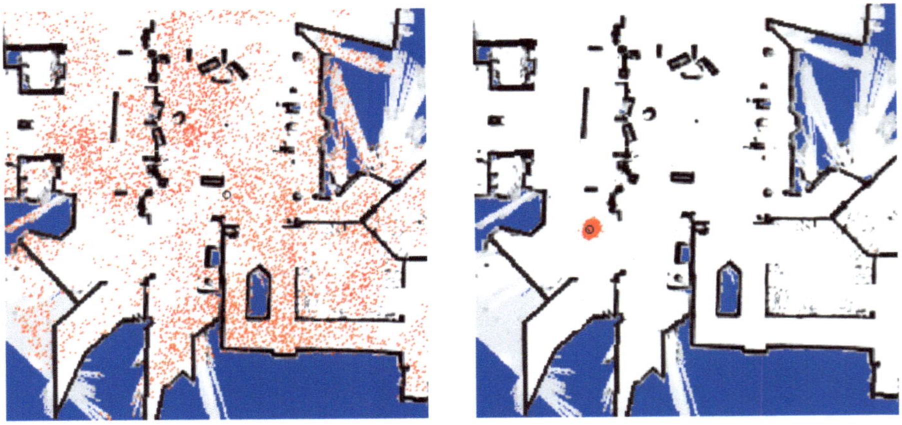

(a) Global localization – After incorporating only a few observations, the pose of the robot is very uncertain.

(b) After incorporating many observations, the particle filter has converged to a unimodal posterior.

Fig. 3.1. Particle filtering for robot localization.

Particle filters have been applied successfully to a variety of real world estimation problems [21, 44, 81]. One of the most common examples of particle filtering in robotics is Monte Carlo Localization, or MCL [89]. In MCL, a set of particles is used to represent the distribution of possible poses of a robot relative to a fixed map. An example is shown in Figure 3.1. In this example, the robot is given no prior information about its pose. This complete uncertainty is represented by scattering particles with uniform probability throughout the map, as shown in Figure 3.1(a). Figure 3.1(b) shows the particle filter after incorporating a number of controls and observations. At this point, the posterior has converged to an approximately unimodal distribution.

The capability to track multi-modal beliefs and include non-linear motion and measurement models makes the performance of particle filters particularly robust. However, the number of particles needed to track a given belief may, in the worst case, scale exponentially with the dimensionality of the state space. As such, standard particle filtering algorithms are restricted to problems of relatively low dimensionality. Particle filters are especially ill-suited to the SLAM problem, which may have millions of dimensions. However, the following sections will show how the SLAM problem can be factored into a set of independent landmark estimation problems conditioned on an estimate of the robot's path. The robot path posterior is of low dimensionality and can be estimated efficiently using a particle filter. The resulting algorithm, called FastSLAM, is an example of a Rao-Blackwellized particle filter [21, 22, 23].

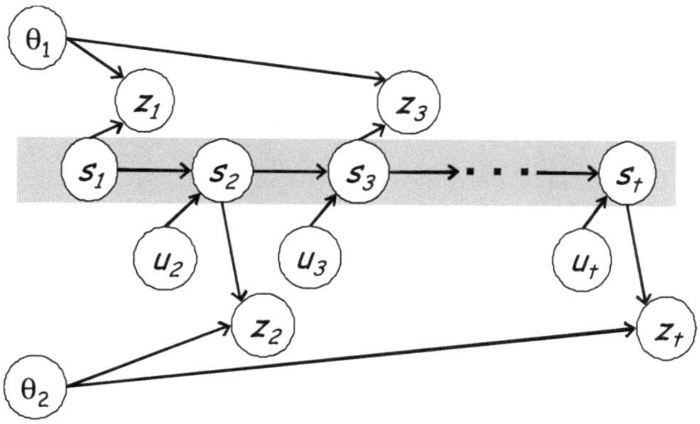

Fig. 3.2. Factoring the SLAM Problem – If the true path of the robot is known (the shaded region), then the positions of the landmarks θ_1 and θ_2 are conditionally independent.

3.2 Factored Posterior Representation

The majority of SLAM approaches are based on estimating the posterior over maps and robot *pose*.

$$p(s_t, \Theta \mid z^t, u^t, n^t) \tag{3.1}$$

FastSLAM computes a slightly different quantity, the posterior over maps and robot *paths*.

$$p(s^t, \Theta \mid z^t, u^t, n^t) \tag{3.2}$$

This subtle difference will allow us to factor the SLAM posterior into a product of simpler terms. Figure 3.2 revisits the interpretation of the SLAM problem as a Dynamic Bayes Network (DBN). In the scenario depicted by the DBN, the robot observes landmark θ_1 at time $t = 1$, θ_2 at time $t = 2$, and then re-observes landmark θ_1 at time $t = 3$. The gray shaded area represents the path of the robot from time $t = 1$ to the present time. From this diagram, it is evident that there are important conditional independences in the SLAM problem. In particular, if the true path of the robot is known, the position of landmark θ_1 is conditionally independent of landmark θ_2. Using the terminology of DBNs, the robot's path "d-separates" the two landmark nodes θ_1 and θ_2. For a complete description of d-separation see [74, 80].

This conditional independence has an important consequence. Given knowledge of the robot's path, an observation of one landmark will not provide any information about the position of any other landmark. In other words, if an oracle told us the true path of the robot, we could estimate the position of every landmark as an independent quantity. This means that the SLAM posterior (3.2) can be factored into a product of simpler terms.

$$p(s^t, \Theta \mid z^t, u^t, n^t) = \underbrace{p(s^t \mid z^t, u^t, n^t)}_{\text{path posterior}} \underbrace{\prod_{n=1}^{N} p(\theta_n \mid s^t, z^t, u^t, n^t)}_{\text{landmark estimators}} \tag{3.3}$$

This factorization, first developed by Murphy [66], states that the SLAM posterior can be separated into a product of a robot path posterior $p(s^t \mid z^t, u^t, n^t)$, and N landmark posteriors conditioned on the robot's path. It is important to note that this factorization is exact; it follows directly from the structure of the SLAM problem.

3.2.1 Proof of the FastSLAM Factorization

The FastSLAM factorization can be derived directly from the SLAM path posterior (3.2). Using the definition of conditional probability, the SLAM posterior can be rewritten as:

$$p(s^t, \Theta \mid z^t, u^t, n^t) = p(s^t \mid z^t, u^t, n^t)\, p(\Theta \mid s^t, z^t, u^t, n^t) \tag{3.4}$$

Thus, to derive the factored posterior (3.3), it suffices to show the following for all non-negative values of t:

$$p(\Theta \mid s^t, z^t, u^t, n^t) = \prod_{n=1}^{N} p(\theta_n \mid s^t, z^t, u^t, n^t) \tag{3.5}$$

Proof of this statement can be demonstrated through induction. Two intermediate results must be derived in order to achieve this result. The first quantity to be derived is the probability of the observed landmark θ_{n_t} conditioned on the data. This quantity can be rewritten using Bayes Rule.

$$p(\theta_{n_t} \mid s^t, z^t, u^t, n^t) \overset{Bayes}{=} \frac{p(z_t \mid \theta_{n_t}, s^t, z^{t-1}, u^t, n^t)}{p(z_t \mid s^t, z^{t-1}, u^t, n^t)}\, p(\theta_{n_t} \mid s^t, z^{t-1}, u^t, n^t) \tag{3.6}$$

Note that the current observation z_t depends solely on the current state of the robot and the landmark being observed. In the rightmost term of (3.6), we similarly notice that the current pose s_t, the current action u_t, and the current data association n_t have no effect on θ_{n_t} without the current observation z_t. Thus, all of these variables can be dropped.

$$p(\theta_{n_t} \mid s^t, z^t, u^t, n^t) \overset{Markov}{=} \frac{p(z_t \mid \theta_{n_t}, s_t, n_t)}{p(z_t \mid s^t, z^{t-1}, u^t, n^t)}\, p(\theta_{n_t} \mid s^{t-1}, z^{t-1}, u^{t-1}, n^{t-1}) \tag{3.7}$$

Next, we solve for the rightmost term of (3.7) to get:

$$p(\theta_{n_t} \mid s^{t-1}, z^{t-1}, u^{t-1}, n^{t-1}) = \frac{p(z_t \mid s^t, z^{t-1}, u^t, n^t)}{p(z_t \mid \theta_{n_t}, s_t, n_t)}\, p(\theta_{n_t} \mid s^t, z^t, u^t, n^t) \tag{3.8}$$

The second intermediate result we need is $p(\theta_{n \neq n_t} \mid s^t, z^t, u^t, n^t)$, the probability of any landmark that is not observed conditioned on the data. This is simple, because the landmark posterior will not change if the landmark is not observed. Thus, the landmark posterior at time t is equal to the posterior at time $t-1$.

$$p(\theta_{n \neq n_t} \mid s^t, z^t, u^t, n^t) \overset{Markov}{=} p(\theta_{n \neq n_t} \mid s^{t-1}, z^{t-1}, u^{t-1}, n^{t-1}) \tag{3.9}$$

With these two intermediate results, we can now perform the proof by induction. First, we assume the following induction hypothesis at time $t-1$.

$$p(\theta \mid s^{t-1}, z^{t-1}, u^{t-1}, n^{t-1}) = \prod_{n=1}^{N} p(\theta_n \mid s^{t-1}, z^{t-1}, u^{t-1}, n^{t-1}) \tag{3.10}$$

For the induction base case of $t = 0$, no observations have been incorporated into the SLAM posterior. Therefore, for $t = 0$ the factorization (3.5) is trivially true.

In general when $t > 0$, we once again use Bayes rule to expand the left side of (3.5).

$$p(\theta \mid s^t, z^t, u^t, n^t) \overset{Bayes}{=} \frac{p(z_t \mid \theta, s^t, z^{t-1}, u^t, n^t)}{p(z_t \mid s^t, z^{t-1}, u^t, n^t)} \, p(\theta \mid s^t, z^{t-1}, u^t, n^t) \tag{3.11}$$

Again, z_t only depends on θ, s_t, and n_t, so the numerator of the first term in (3.11) can be simplified. The landmark position θ does not depend on s_t, u_t, or n_t, without the current observation z_t, so the second term can also be simplified.

$$\overset{Markov}{=} \frac{p(z_t \mid \theta_{n_t}, s_t, n_t)}{p(z_t \mid s^t, z^{t-1}, u^t, n^t)} \, p(\theta \mid s^{t-1}, z^{t-1}, u^{t-1}, n^{t-1}) \tag{3.12}$$

Now the rightmost term in (3.12) can be replaced with the induction hypothesis (refeq:induction-hypothesis).

$$\overset{Induction}{=} \frac{p(z_t \mid \theta_{n_t}, s_t, n_t)}{p(z_t \mid s^t, z^{t-1}, u^t, n^t)} \prod_{n=1}^{N} p(\theta_n \mid s^{t-1}, z^{t-1}, u^{t-1}, n^{t-1}) \tag{3.13}$$

Replacing the terms of the product with the two intermediate results (3.8) and (3.9), we get:

$$p(\theta \mid s^t, z^t, u^t, n^t) = p(\theta_{n_t} \mid s^t, z^t, u^t, n^t) \prod_{n \neq n_t}^{N} p(\theta_i \mid s^t, z^t, u^t, n^t) \tag{3.14}$$

which is equal to the product of the individual landmark posteriors (3.5):

$$p(\theta \mid s^t, z^t, u^t, n^t) = \prod_{n=1}^{N} p(\theta_n \mid s^t, z^t, u^t, n^t) \tag{3.15}$$

qed.

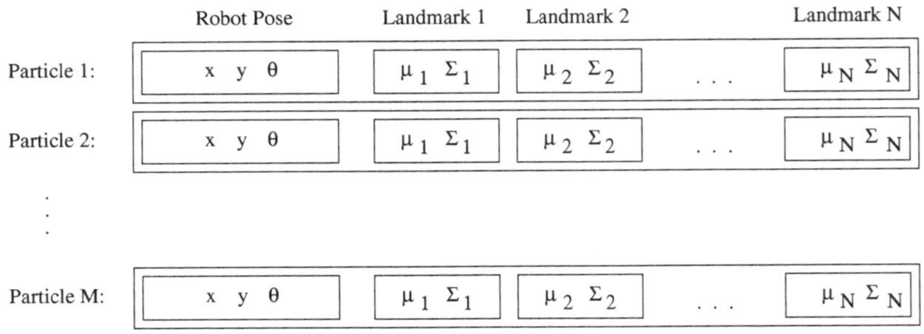

Fig. 3.3. There are M particles in the particle filter. Each particle contains N independent EKFs. No explicit cross-correlations are maintained between the landmark estimates.

3.3 The FastSLAM 1.0 Algorithm

The factorization of the posterior (3.3) highlights important structure in the SLAM problem that is ignored by SLAM algorithms that estimate an unstructured posterior. This structure suggests that under the appropriate conditioning, no cross-correlations between landmarks have to be maintained explicitly. FastSLAM exploits the factored representation by maintaining $N+1$ filters, one for each term in (3.3). By doing so, all $N+1$ filters are low-dimensional.

FastSLAM estimates the first term in (3.3), the robot path posterior, using a particle filter. The remaining N conditional landmark posteriors $p(\theta_n \mid s^t, z^t, u^t, n^t)$ are estimated using EKFs. Each EKF tracks a single landmark position, and therefore is low-dimensional and fixed in size. The landmark EKFs are all conditioned on robot paths, with each particle in the particle filter possessing its own set of EKFs. In total, there are $N \cdot M$ EKFs, where M is the total number of particles in the particle filter. The particle filter is depicted graphically in Figure 3.3. Each FastSLAM particle is of the form:

$$S_t^{[m]} = \langle s^{t,[m]}, \mu_{1,t}^{[m]}, \Sigma_{1,t}^{[m]}, \ldots, \mu_{N,t}^{[m]}, \Sigma_{N,t}^{[m]} \rangle \tag{3.16}$$

The bracketed notation $[m]$ indicates the index of the particle; $s^{t,[m]}$ is the m-th particle's path estimate, and $\mu_{n,t}^{[m]}$ and $\Sigma_{n,t}^{[m]}$ are the mean and covariance of the Gaussian representing the n-th feature location conditioned on the path $s^{t,[m]}$. Together all of these quantities form the m-th particle $S_t^{[m]}$, of which there are a total of M in the FastSLAM posterior. Filtering, that is, calculating the posterior at time t from the one at time $t-1$, involves generating a new particle set S_t from S_{t-1}, the particle set one time step earlier. The new particle set incorporates the latest control u_t and measurement z_t (with corresponding data association n_t). This update is performed in four steps.

First, a new robot pose is drawn for each particle that incorporates the latest control. Each pose is added to the appropriate robot path estimate $s^{t-1,[m]}$. Next, the landmark EKFs corresponding to the observed landmark

Table 3.1. Outline of the basic FastSLAM algorithm

```
1. Sample a new robot pose for each particle given the new control
2. Update the landmark EKFs of the observed feature in each
particle
3. Calculate an importance weight for each particle
4. Draw an new, unweighted particle set using importance resampling
```

are updated with the new observation. Since the robot path particles are not drawn from the true path posterior, each particle is given an importance weight to reflect this difference. A new set of particles S_t is drawn from the weighted particle set using importance resampling. This importance resampling step is necessary to insure that the particles are distributed according to the true posterior (in the limit of infinite particles). The four basic steps of the FastSLAM algorithm [59], shown in Table 3.1, will be explained in detail in the following four sections.

3.3.1 Sampling a New Pose

The particle set S_t is calculated incrementally, from the set S_{t-1} at time $t-1$, the observation z_t, and the control u_t. Since we cannot draw samples directly from the SLAM posterior at time t, we will instead draw samples from a simpler distribution called the proposal distribution, and correct for the difference using a technique called importance sampling.

In general, importance sampling is an algorithm for drawing samples from functions for which no direct sampling procedure exists [55]. Each sample drawn from the proposal distribution is given a weight equal to the ratio of the posterior distribution to the proposal distribution at that point in the sample space. A new set of unweighted samples is drawn from the weighted set with probabilities in proportion to the weights. This process is an instantiation of Rubin's Sampling Importance Resampling (SIR) algorithm [79].

The proposal distribution of FastSLAM generate guesses of the robot's pose at time t given each particle $S_{t-1}^{[m]}$. This guess is obtained by sampling from the probabilistic motion model.

$$s_t^{[m]} \sim p(s_t \mid u_t, s_{t-1}^{[m]}) \tag{3.17}$$

This estimate is added to a temporary set of particles, along with the path $s^{t-1,[m]}$. Under the assumption that the set of particles S_{t-1} is distributed according to $p(s^{t-1} \mid z^{t-1}, u^{t-1}, n^{t-1})$, which is asymptotically correct, the new particles drawn from the proposal distribution are distributed according to:

$$p(s^t \mid z^{t-1}, u^t, n^{t-1}) \tag{3.18}$$

It is important to note that the motion model can be any non-linear function. This is in contrast to the EKF, which requires the motion model to be

Fig. 3.4. Samples drawn from the probabilistic motion model.

linearized. The only practical limitation on the measurement model is that samples can be drawn from it conveniently. Regardless of the proposal distribution, drawing a new pose is a constant-time operation for every particle. It does not depend on the size of the map.

A simple four parameter motion model was used for all of the planar robot experiments in this book. This model assumes that the velocity of the robot is constant over the time interval covered by each control. Each control u_t is two-dimensional and can be written as a translational velocity v_t and a rotational velocity ω_t. The model further assumes that the error in the controls is Gaussian. Note that this does not imply that error in the robot's motion will also be Gaussian; the robot's motion is a non-linear function of the controls and the control noise.

The errors in translational and rotational velocity have an additive and a multiplicative component. Throughout this book, the notation $\mathcal{N}(x; \mu, \Sigma)$ will be used to denote a normal distribution over the variable x with mean μ and covariance Σ.

$$v_t^{'} \sim \mathcal{N}(v_t, \alpha_1 v_t + \alpha_2) \tag{3.19}$$

$$\omega_t^{'} \sim \mathcal{N}(\omega_t, \alpha_3 \omega_t + \alpha_4) \tag{3.20}$$

This motion model is able to represent the slip and skid errors errors that occur in typical ground vehicles [8]. The first step to drawing a new robot pose from this model is to draw a new translational and rotational velocity according to the observed control. The new pose s_t can be calculated by simulating the new control forward from the previous pose $s_{t-1}^{[m]}$. Figure 3.4 shows 250 samples drawn from this motion model given a curved trajectory. In this simulated example, the translational error of the robot is low, while the rotational error is high.

3.3.2 Updating the Landmark Estimates

FastSLAM represents the conditional landmark estimates $p(\theta_n \mid s^t, z^t, u^t, n^t)$ in (3.3) using low-dimensional EKFs. For now, I will assume that the data associations n^t are known. In section 3.4, this restriction will be removed.

Since the landmark estimates are conditioned on the robot's path, N EKFs are attached to each particle in S_t. The posterior over the n-th landmark position θ_n is easily obtained. Its computation depends on whether $n = n_t$, that is, whether or not landmark θ_n was observed at time t. For the observed landmark θ_{n_t}, we follow the usual procedure of expanding the posterior using Bayes Rule.

$$p(\theta_{n_t} \mid s^t, z^t, u^t, n^t) \overset{Bayes}{=} \eta\, p(z_t \mid \theta_{n_t}, s^t, z^{t-1}, u^t, n^t)\, p(\theta_{n_t} \mid s^t, z^{t-1}, u^t, n^t) \tag{3.21}$$

Next, the Markov property is used to simplify both terms of the equation. The observation z_t only depends on θ_{n_t}, s_t, and n_t. Similarly, θ_{n_t} is not affected by s_t, u_t, or n_t without the observation z_t.

$$p(\theta_{n_t} \mid s^t, z^t, u^t, n^t) \overset{Markov}{=} \eta\, p(z_t \mid \theta_{n_t}, s_t, n_t)\, p(\theta_{n_t} \mid s^{t-1}, z^{t-1}, u^{t-1}, n^{t-1}) \tag{3.22}$$

For $n \neq n_t$, we leave the landmark posterior unchanged.

$$p(\theta_{n \neq n_t} \mid s^t, z^t, u^t, n^t) = p(\theta_{n \neq n_t} \mid s^{t-1}, z^{t-1}, u^{t-1}, n^{t-1}) \tag{3.23}$$

FastSLAM implements the update equation (3.22) using an EKF. As in EKF solutions to SLAM, this filter uses a linear Gaussian approximation for the perceptual model. We note that, with an actual linear Gaussian observation model, the resulting distribution $p(\theta_n \mid s^t, z^t, u^t, n^t)$ is exactly Gaussian, even if the motion model is non-linear. This is a consequence of sampling over the robot's pose.

The non-linear measurement model $g(s_t, \theta_{n_t})$ will be approximated using a first-order Taylor expansion. The landmark estimator is conditioned on a fixed robot path, so this expansion is only over θ_{n_t}. We will assume that measurement noise is Gaussian with covariance R_t.

$$\hat{z}_t = g(s_t^{[m]}, \mu_{n_t,t-1}) \tag{3.24}$$

$$G_{\theta_{n_t}} = \nabla_{\theta_{n_t}} g(s_t, \theta_{n_t})\big|_{s_t = s_t^{[m]};\, \theta_{n_t} = \mu_{n_t,t-1}^{[m]}} \tag{3.25}$$

$$g(s_t, \theta_{n_t}) \approx \hat{z}_t + G_\theta(\theta_{n_t} - \mu_{n_t,t-1}^{[m]}) \tag{3.26}$$

Under this approximation, the first term of the product (3.22) is distributed as follows:

$$p(z_t \mid \theta_i, s_t, n_t) \sim \mathcal{N}(z_t; \hat{z}_t + G_\theta(\theta_{n_t} - \mu_{n_t,t-1}^{[m]}), R_t) \tag{3.27}$$

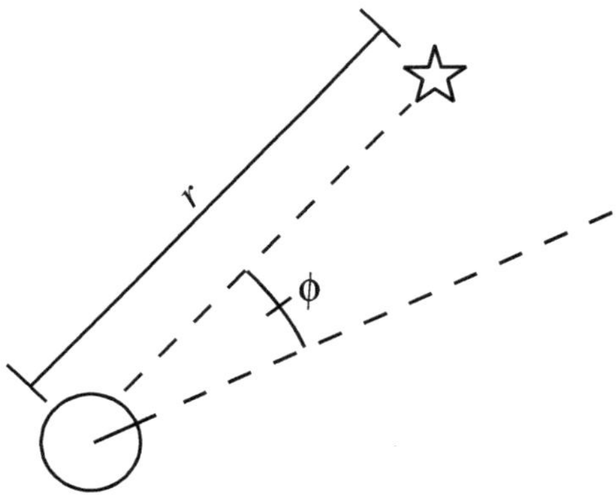

Fig. 3.5. Robot observing the range r and bearing ϕ to a landmark.

The second term of the product in (3.22) is also a Gaussian, equal to the state of the EKF at time $t-1$.

$$p(\theta_i \mid s^{t-1}, z^{t-1}, u^{t-1}, n^{t-1}) \sim \mathcal{N}(\theta_{n_t}; \mu^{[m]}_{n_t,t-1}, \Sigma^{[m]}_{n_t,t-1}) \tag{3.28}$$

The mean and covariance of the product can be obtained using the standard EKF update equations [3].

$$\hat{z}_t = g(s^{[m]}_t, \mu_{n_t,t-1}) \tag{3.29}$$

$$G_{\theta_{n_t}} = \nabla_{\theta_{n_t}} g(s_t, \theta_{n_t})\big|_{s_t=s^{[m]}_t;\theta_{n_t}=\mu^{[m]}_{n_t,t-1}} \tag{3.30}$$

$$Z_{n,t} = G_{\theta_{n_t}} \Sigma^{[m]}_{n_t,t-1} G^T_{\theta_{n_t}} + R_t \tag{3.31}$$

$$K_t = \Sigma^{[m]}_{n_t,t-1} G^T_{\theta_{n_t}} Z^{-1}_{n,t} \tag{3.32}$$

$$\mu^{[m]}_{n_t,t} = \mu^{[m]}_{n_t,t-1} + K_t(z_t - \hat{z}_t) \tag{3.33}$$

$$\Sigma^{[m]}_{n_t,t} = (I - K_t G_{\theta_{n_t}})\Sigma^{[m]}_{n_t,t-1} \tag{3.34}$$

Updating the landmark filters is a constant-time operation per particle because each landmark filter is of constant size. The amount of time necessary to incorporate an observation does not depend on the total number of landmarks.

In the planar SLAM case with fully observable landmarks, the robot commonly observes the range and bearing to nearby landmarks as shown in Figure 3.5. Assuming that the robot pose s_t is described as $\langle s_{t,x}, s_{t,y}, s_{t,\theta}\rangle$ and the current landmark position is written as $\langle \theta_{n_t,x}, \theta_{n_t,y}\rangle$, the measurement function $g(s_t, \theta_{n_t})$ can be written as the following matrix function:

$$g(s_t, \theta_{n_t}) = \begin{bmatrix} r(s_t, \theta_{n_t}) \\ \phi(s_t, \theta_{n_t}) \end{bmatrix} = \begin{bmatrix} \sqrt{(\theta_{n_t,x} - s_{t,x})^2 + (\theta_{n_t,y} - s_{t,y})^2} \\ \tan^{-1}\left(\frac{\theta_{n_t,y} - s_{t,y}}{\theta_{n_t,x} - s_{t,x}}\right) - s_{t,\theta} \end{bmatrix} \tag{3.35}$$

The Jacobian $G_{\theta_{n_t}}$ is then equal to:

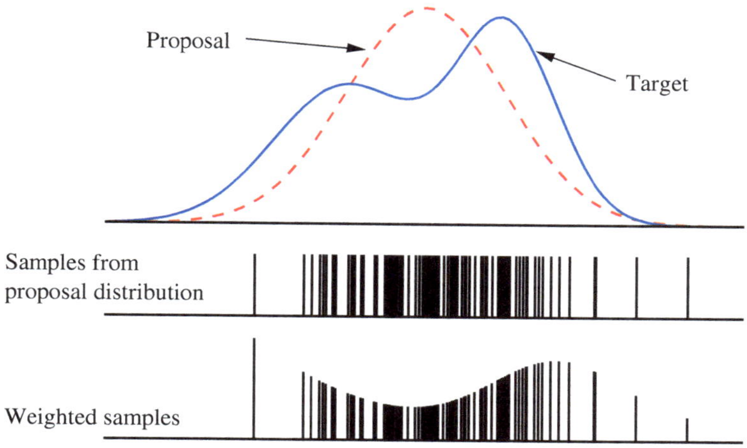

Fig. 3.6. Samples cannot be drawn conveniently from the target target distribution (shown as a solid line). Instead, the importance sampler draws samples from the proposal distribution (dashed line), which has a simpler form. Below, samples drawn from the proposal distribution are drawn with lengths proportional to their importance weights.

$$G_{\theta_{n_t}} = \begin{bmatrix} \frac{\theta_{n_t,x}-s_{t,x}}{\sqrt{q}} & \frac{\theta_{n_t,y}-s_{t,y}}{\sqrt{q}} \\ -\frac{\theta_{n_t,y}-s_{t,y}}{q} & \frac{\theta_{n_t,x}-s_{t,x}}{q} \end{bmatrix} \tag{3.36}$$
$$q = (\theta_{n_t,x} - s_{t,x})^2 + (\theta_{n_t,y} - s_{t,y})^2$$

3.3.3 Calculating Importance Weights

Samples from the proposal distribution are distributed according to $p(s^t \mid z^{t-1}, u^t, n^{t-1})$, and therefore do not match the desired posterior $p(s^t \mid z^t, u^t, n^t)$. This difference is corrected through importance sampling. An example of importance sampling is shown in Figure 3.6. Instead of sampling directly from the target distribution (shown as a solid line), samples are drawn from a simpler proposal distribution, a Gaussian (shown as a dashed line). In regions where the target distribution is larger than the proposal distribution, the samples receive higher weights. As a result, samples in this region will be picked more often. In regions where the target distribution is smaller than the proposal distribution, the samples will be given lower weights. In the limit of infinite samples, this procedure will produce samples distributed according to the target distribution.

For FastSLAM, the importance weight of each particle $w_t^{[i]}$ is equal to the ratio of the SLAM posterior and the proposal distribution described previously.

$$w_t^{[m]} = \frac{\text{target distribution}}{\text{proposal distribution}} = \frac{p(s^{t,[m]} \mid z^t, u^t, n^t)}{p(s^{t,[m]} \mid z^{t-1}, u^t, n^{t-1})} \tag{3.37}$$

The numerator of (3.37) can be expanded using Bayes Rule. The normalizing constant in Bayes Rule can be safely ignored because the particle weights will be normalized before resampling.

$$w_t^{[m]} \overset{Bayes}{\propto} \frac{p(z_t \mid s^{t,[m]}, z^{t-1}, u^t, n^t) p(s^{t,[m]} \mid z^{t-1}, u^t, n^t)}{p(s^{t,[m]} \mid z^{t-1}, u^t, n^{t-1})} \tag{3.38}$$

The second term of the numerator is not conditioned on the latest observation z_t, so the data association n_t cannot provide any information about the robot's path. Therefore it can be dropped.

$$\begin{aligned} w_t^{[m]} &\overset{Markov}{=} \frac{p(z_t \mid s^{t,[m]}, z^{t-1}, u^t, n^t) p(s^{t,[m]} \mid z^{t-1}, u^t, n^{t-1})}{p(s^{t,[m]} \mid z^{t-1}, u^t, n^{t-1})} \\ &= p(z_t \mid s^{t,[m]}, z^{t-1}, u^t, n^t) \end{aligned} \tag{3.39}$$

The landmark estimator is an EKF, so this observation likelihood can be computed in closed form. This probability is commonly computed in terms of "innovation," or the difference between the actual observation z_t and the predicted observation $\hat{z}_t$. The sequence of innovations in the EKF is Gaussian with zero mean and covariance $Z_{n_t,t}$, where $Z_{n_t,t}$ is the innovation covariance matrix defined in (3.31) [3]. The probability of the observation z_t is equal to the probability of the innovation $z_t - \hat{z}_t$ being generated by this Gaussian, which can be written as:

$$w_t^{[m]} = \frac{1}{\sqrt{|2\pi Z_{n_t,t}|}} \exp\{-\frac{1}{2}(z_t - \hat{z}_{n_t,t})^T [Z_{n_t,t}]^{-1} (z_t - \hat{z}_{n_t,t})\} \tag{3.40}$$

Calculating the importance weight is a constant-time operation per particle. This calculation depends only on the dimensionality of the observation, which is constant for a given application.

3.3.4 Importance Resampling

Once the temporary particles have been assigned weights, a new set of samples S_t is drawn from this set with replacement, with probabilities in proportion to the weights. A variety of sampling techniques for drawing S_t can be found in [9]. In particular, Madow's systematic sampling algorithm [56] is simple to implement and produces accurate results.

Implemented naively, resampling requires time linear in the number of landmarks N. This is due to the fact that each particle must be copied to the new particle set, and the length of each particle is proportional to N. In general, only a small fraction of the total landmarks will be observed at any one time, so copying the entire particle can be quite inefficient. In Section 3.7, we will show how a more sophisticated particle representation can eliminate unnecessary copying and reduce the computational requirement of FastSLAM to $O(M \log N)$.

3.3.5 Robot Path Posterior Revisited

At first glace, factoring the SLAM problem using the path of the robot may seem like a bad idea, because the length of the FastSLAM particles will grow over time. However, none of the the FastSLAM update equations depend on the total path length t. In fact, only the most recent pose $s_{t-1}^{[m]}$ is used to update the particle set. Consequently, we can silently "forget" all but the most recent robot pose in the parameterization of each particle. This avoids the obvious computational problem that would result if the dimensionality of the particle filter grows over time.

3.4 FastSLAM with Unknown Data Association

The biggest limitation of the FastSLAM algorithm described thus far is the assumption that the data associations n^t are known. In practice, this is rarely the case. This section extends the FastSLAM algorithm to domains in which the mapping between observations and landmarks is not known [57]. The classical solution to the data association problem in SLAM is to chose n_t such that it maximizes the likelihood of the sensor measurement z_t given all available data [18].

$$\hat{n}_t = \underset{n_t}{\operatorname{argmax}}\, p(z_t \mid n_t, \hat{n}^{t-1}, s^t, z^{t-1}, u^t) \tag{3.41}$$

The term $p(z_t \mid n_t, \hat{n}^{t-1}, s^t, z^{t-1}, u^t)$ is referred to as a *likelihood,* and this approach is an example of a *maximum likelihood* (ML) estimator. ML data association is also called "nearest neighbor" data association, interpreting the negative log likelihood as a distance function. For Gaussians, the negative log likelihood is Mahalanobis distance, and the estimator selects data associations by minimizing this Mahalanobis distance.

In the EKF-based SLAM approaches described in Chapter 2, a single data association is chosen for the entire filter. As a result, these algorithms tend to be brittle to failures in data association. A single data association error can induce significant errors in the map, which in turn cause new data association errors, often with fatal consequences. A better understanding of how uncertainty in the SLAM posterior generates data association ambiguity will demonstrate how simple data association heuristics often fail.

3.4.1 Data Association Uncertainty

Two factors contribute to uncertainty in the SLAM posterior: measurement noise and motion noise. As measurement noise increases, the distributions of possible observations of every landmark become more uncertain. If measurement noise is sufficiently high, the distributions of observations from nearby landmarks will begin to overlap substantially. This overlap leads to ambiguity

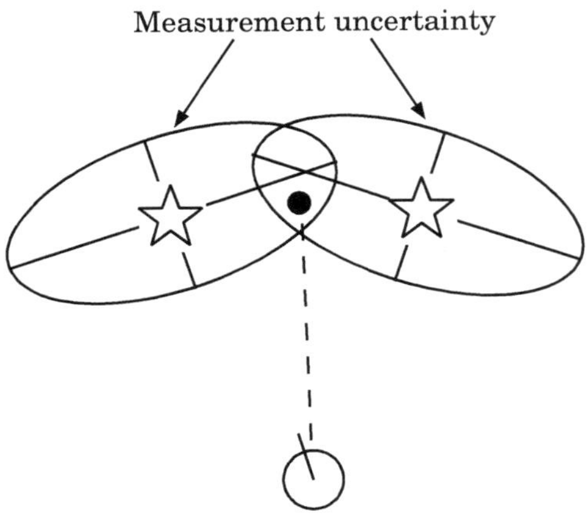

Fig. 3.7. Measurement ambiguity – high measurement error leads to ambiguity between nearby landmarks.

in the identity of the landmarks. We will refer to data association ambiguity caused by measurement noise as *measurement ambiguity.* An example of measurement ambiguity is shown in Figure 3.7. The two ellipses depict the range of probable observations from two different landmarks. The observation, shown as an black circle, plausibly could have come from either landmark.

Attributing an observation to the wrong landmark due to measurement ambiguity will increase the error of the map and robot pose, but its impact will be relatively minor. Since the observation could have been generated by either landmark with high probability, the effect of the observation on the landmark positions and the robot pose will be small. The covariance of one landmark will be slightly overestimated, while the covariance of the second will be slightly underestimated. If multiple observations are incorporated per control, a data association mistake due to measurement ambiguity of one observation will have relatively little impact on the data association decisions for the other observations.

Ambiguity in data association caused by motion noise can have much more severe consequences on estimation accuracy. Higher motion noise will lead to higher pose uncertainty after incorporating a control. If this pose uncertainty is high enough, assuming different robot poses in this distribution will imply drastically different ML data association hypotheses for the subsequent observations. This motion ambiguity, shown in Figure 3.8, is easily induced if there is significant rotational error in the robot's motion. Moreover, if multiple observations are incorporated per control, the pose of the robot will correlate the data association decisions of all of the observations. If the SLAM algorithm chooses the wrong data association for a single observation due to motion ambiguity, the rest of the data associations also will be wrong with high probability.

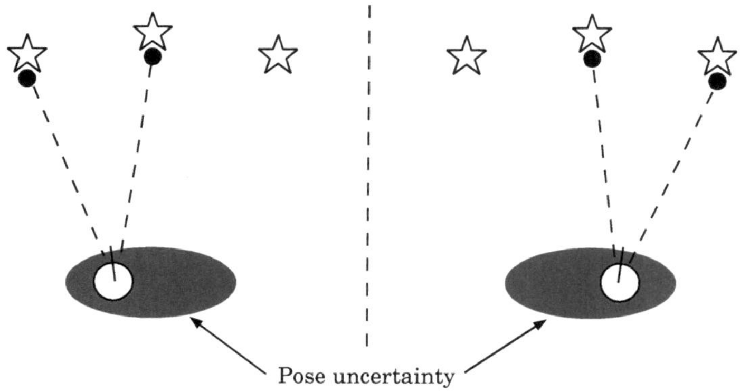

Fig. 3.8. Motion ambiguity-uncertainty due to the motion of the robot may result in significantly different data association hypotheses for different robot poses. Also, motion error correlates the data associations of simultaneous observations.

3.4.2 Per-Particle Data Association

Unlike most EKF-based SLAM algorithms, FastSLAM takes a multi-hypothesis approach to the data association problem. Each particle represents a different hypothesized path of the robot, so data association decisions can be made on a per-particle basis. Particles that pick the correct data association will receive high weights because they explain the observations well. Particles that pick wrong associations will receive low weights and be removed in a future resampling step.

Per-particle data association has several important advantages over standard ML data association. First, it factors robot pose uncertainty out of the data association problem. Since motion ambiguity is the more severe form of data association ambiguity, conditioning the data association decisions on hypothesized robot paths seems like a logical choice. Given the scenario in Figure 3.8, some of the particles would draw new robot poses consistent with data association hypothesis on the left, while others would draw poses consistent with the data association hypothesis on the right.

Doing data association on a per-particle basis also makes the data association problem easier. In the EKF, the uncertainty of a landmark position is due to both uncertainty in the pose of the robot and measurement error. In FastSLAM, uncertainty of the robot pose is represented by the entire particle set. The landmark filters in a single particle are not affected by motion noise because they are conditioned on a specific robot path. This is especially useful if the robot has noisy motion and an accurate sensor.

Another consequence of per-particle data association is implicit, delayed-decision making. At any given time, some fraction of the particles will receive plausible, yet wrong, data associations. In the future, the robot may receive a new observation that clearly refutes these previous assignments. At this

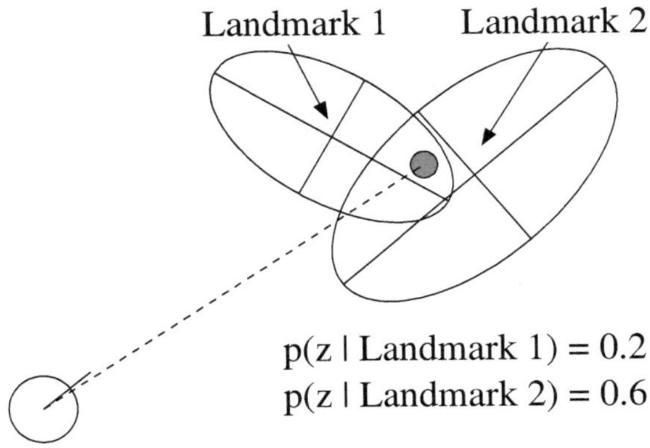

Fig. 3.9. Ambiguous Data Association.

point, the particles with wrong data associations will receive low weight and likely be removed from the filter. As a result of this process, the effect of a wrong data association decision made in the past can be removed from the filter. Moreover, no heuristics are needed in order to remove incorrect old associations from the filter. This is done in a statistically valid manner, simply as a consequence of the resampling step.

Per-Particle Maximum Likelihood Data Association

The simplest approach to per-particle data association is to apply the maximum likelihood (ML) data association heuristic (3.41), only on a per-particle basis. Since the landmark estimators are EKFs, the likelihood in (3.41) can be calculated using innovations. This likelihood is exactly the same as the importance weight calculated in (3.40) for the original FastSLAM algorithm. If the value of this likelihood falls below some threshold p_0, a new landmark is added to the particle.

$$p(z_t \mid s^{t,[m]}, z^{t-1}, u^t, n^{t-1}) = \frac{1}{\sqrt{|2\pi Z_{n,t}|}} \exp\{-\frac{1}{2}(z_t - \hat{z}_{n,t})^T [Z_{n,t}]^{-1} (z_t - \hat{z}_{n,t})\} \tag{3.42}$$

Monte Carlo Data Association

While per-particle ML data association addresses motion ambiguity, it does not address measurement ambiguity. Each observation is paired with the landmark most likely to have generated it, however if measurement error is high there might be several plausible data associations per observation. Another approach to data association is to assign the correspondences probabilistically in accordance to their likelihoods. This approach can be described as *Monte Carlo Data Association.*

Figure 3.9 shows a robot making an observation that could have been generated by either one of two landmarks. The probability of the observation being generated by landmark 1 is 0.2, and 0.6 for landmark 2. The Monte Carlo Data Association procedure would assign the data association $n_t = 2$ exactly 75% of the time (since $0.6/(0.2 + 0.6) = 0.25$), and it would assign $n_t = 1$ the remaining 25% of the time (since $0.2/(0.2 + 0.6) = 0.75$). In general, the ML data association scheme will pick the correct data association more often than the Monte Carlo scheme. However, ML will never consider the possibility that nearby data associations may have been exchanged due to measurement ambiguity. As Monte Carlo will generate exponentially more possible data associations as measurement error increases, this scheme will require a larger number of particles, in general, to maintain the same accuracy.

A third possible data association scheme for FastSLAM was proposed by Niento et al. in [71]. Their procedure enumerates all K plausible data associations for a given observation and particle. Plausible data associations are defined as correspondences with a likelihood above some minimum threshold. Each particle is then duplicated K times, and the observation is incorporated into each particle with the corresponding data association. Later, the particle set is reduced back to M particles by the resampling process. Since each particle will survive with probability proportional to the likelihood of the observation, this scheme is equivalent to the Monte Carlo data association procedure.

3.4.3 Adding New Landmarks

Adding a new landmark to FastSLAM can be difficult decision to make, just as with EKF-based algorithms. This is especially true when an individual measurement is insufficient to constrain the new landmark in all dimensions [13]. If the measurement function $g(\theta_{n_t}, s_t)$ is invertible, however, a single measurement is sufficient to initialize a new landmark. Each observation defines a Gaussian:

$$\mathcal{N}(z_t; \hat{z}_t + G_{\theta_{n_t}}(\theta_{n_t} - \mu_{n_t,t}^{[m]}), R_t) \tag{3.43}$$

This Gaussian can be written explicitly as:

$$\frac{1}{\sqrt{|2\pi R_t|}} \exp\left\{ -\frac{1}{2}(z_t - \hat{z}_t - G_{\theta_{n_t}}(\theta_{n_t} - \mu_{n_t,t-1}^{[m]}))^T \right. \\ \left. R_t^{-1}(z_t - \hat{z}_t - G_{\theta_{n_t}}(\theta_{n_t} - \mu_{n_t,t-1}^{[m]})) \right\} \tag{3.44}$$

We define a function J to be equal to the negative of the exponent of this Gaussian:

$$J = \frac{1}{2}(z_t - \hat{z}_t - G_{\theta_{n_t}}(\theta_{n_t} - \mu_{n_t,t-1}^{[m]}))^T R_t^{-1}(z_t - \hat{z}_t - G_{\theta_{n_t}}(\theta_{n_t} - \mu_{n_t,t-1}^{[m]})) \tag{3.45}$$

The second derivative of J with respect to θ_{n_t} will be the inverse of the covariance matrix of the Gaussian in landmark coordinates.

$$\frac{\partial J}{\partial \theta_{n_t}} = -(z_t - \hat{z}_t - G_{\theta_{n_t}}(\theta_{n_t} - \mu^{[m]}_{n_t,t-1}))^T R_t^{-1} G_{\theta_{n_t}} \tag{3.46}$$

$$\frac{\partial^2 J}{\partial \theta^2_{n_t}} = G^T_{\theta_{n_t}} R_t^{-1} G_{\theta_{n_t}} \tag{3.47}$$

Consequently, an invertible observation can be used to create a new landmark as follows.

$$\mu^{[m]}_{n_t,t} = g^{-1}(s^{[m]}_t, z_t) \tag{3.48}$$

$$\Sigma^{[m]}_{n_t,t} = \left(G^T_{\theta_{n_t},t} R^{-1} G_{\theta_{n_t},t} \right)^{-1} \tag{3.49}$$

$$w^{[m]}_t = p_0 \tag{3.50}$$

In practice, a simpler initialization procedure also works well. Instead of computing the correct initial covariance, the covariance can be computed by setting the variance of each landmark parameter to a large initial value K and incorporating the first observation. Higher values of K lead to closer approximations of the true covariance, but can also lead to numerical instability.

$$\mu^{[m]}_{n_t,t} = g^{-1}(s^{[m]}_t, z_t) \tag{3.51}$$

$$\Sigma^{[m]}_{n_t,t} = K \cdot I \tag{3.52}$$

Initialization techniques for situations in which g is not invertible (e.g. bearings-only SLAM) are discussed in [12, 13]. These situations require the accumulation of multiple observations in order to estimate the location of a landmark. FastSLAM is currently being applied to the problem of bearings-only SLAM [83].

3.5 Summary of the FastSLAM Algorithm

Table 3.2 summarizes the FastSLAM algorithm with unknown data association. Particles in the complete FastSLAM algorithm have the form:

$$S^{[m]}_t = \langle s^{[m]}_t \mid N^{[m]}_t, \mu^{[m]}_{1,t}, \Sigma^{[m]}_{1,t}, \ldots, \mu^{[m]}_{N^{[m]}_t,t}, \Sigma^{[m]}_{N^{[m]}_t,t} \rangle \tag{3.53}$$

In addition to the latest robot pose $s^{[m]}_t$ and the feature estimates $\mu^{[m]}_{n,t}$ and $\Sigma^{[m]}_{n,t}$, each particle maintains the number of features $N^{[m]}_t$ in its local map. It is interesting to note that each particle may have a different number of landmarks. This is an expressive representation, but it can lead to difficulties in determining the most probable map.

Table 3.2. The FastSLAM 1.0 algorithm.

Algorithm FastSLAM 1.0(S_{t-1}, z_t, R_t, u_t)

$S_t = S_{aux} = \emptyset$

for $m = 1$ **to** M // loop: all particles

  retrieve m-th particle

$$\left\langle s_{t-1}^{[m]}, N_{t-1}^{[m]}, \mu_{1,t-1}^{[m]}, \Sigma_{1,t-1}^{[m]}, \ldots, \mu_{N_{t-1}^{[m]},t-1}^{[m]}, \Sigma_{N_{t-1}^{[m]},t-1}^{[m]} \right\rangle \text{ from } S_{t-1}$$

  draw $s_t^{[m]} \sim p(s_t \mid s_{t-1}^{[m]}, u_t)$ // sample new pose

  for $n = 1$ **to** $N_{t-1}^{[m]}$ // loop over potential data associations

    $G_{\theta,n} = \nabla_{\theta_n} g(\theta_n, s_t)|_{\theta_n = \mu_{n,t-1}^{[i]}; s_t = s_t^{[i]}}$

    $\hat{z}_{n,t} = g(s_t^{[m]}, \mu_{n,t-1}^{[m]})$

    $Z_{n,t} = G_{\theta,n} \Sigma_{n,t-1}^{[m]} G_{\theta,n}^T + R_t$

    $p_{n,t}^{[m]} = |2\pi Z_{n,t}|^{-\frac{1}{2}} \exp\{-\frac{1}{2}(z_t - \hat{z}_{n,t})^T Z_{n,t}^{-1}(z_t - \hat{z}_{n,t})\}$

  end for

  $p_{N_{t-1}^{[m]}+1,t}^{[m]} = p_0$

  $\hat{n}_t = \operatorname{argmax}_n p_{n,t}^{[m]}$ or draw $\hat{n}_t$ with prob. $\propto p_{n,t}^{[m]}$ // pick data association

  if $\hat{n}_t = N_{t-1}^{[m]} + 1$ // is new feature?

    $N_t^{[m]} = N_{t-1}^{[m]} + 1$

    $\mu_{\hat{n}_t,t}^{[m]} = g^{-1}(s_t^{[m]}, \hat{z}_{\hat{n}_t,t})$

    $\Sigma_{\hat{n}_t,t}^{[m]} = \left(G_{\theta,\hat{n}_t}^T R^{-1} G_{\theta,\hat{n}_t}\right)^{-1}$

  else // or is a known feature?

    $N_t^{[m]} = N_{t-1}^{[m]}$

    $K_{\hat{n}_t,t} = \Sigma_{\hat{n}_t,t-1} G_{\theta,\hat{n}_t}^T Z_{\hat{n}_t,t}^{-1}$

    $\mu_{\hat{n}_t,t}^{[m]} = \mu_{\hat{n}_t,t-1}^{[m]} + K_{\hat{n}_t,t}(z_t - \hat{z}_{\hat{n}_t,t})$

    $\Sigma_{\hat{n}_t,t}^{[m]} = (I - K_{\hat{n}_t,t} G_{\theta,\hat{n}_t}) \Sigma_{\hat{n}_t,t-1}^{[m]}$

  end if

  for $n = 1$ **to** $N_t^{[m]}$ do // unobserved features

    if $n \neq \hat{n}_t$

      $\mu_{\theta_n,t}^{[m]} = \mu_{\theta_n,t-1}^{[m]}$

      $\Sigma_{\theta_n,t}^{[m]} = \Sigma_{\theta_n,t-1}^{[m]}$

    end if

  end for

  $w_t^{[m]} = p_{\hat{n}_t,t}^{[m]}$ // save weighted particle

$$\text{add } \left\langle s_t^{[m]}, N_t^{[m]}, \mu_{1,t}^{[m]}, \Sigma_{1,t}^{[m]}, \ldots, \mu_{N_t^{[m]},t}^{[m]}, \Sigma_{N_t^{[m]},t}^{[m]}, w_t^{[m]} \right\rangle \text{ to } S_{aux}$$

end for

for $m = 1$ **to** M // resample

  draw random particle from S_{aux} with probability $\propto w_t^{[m]}$

  add new particle to S_t

end for

return S_t

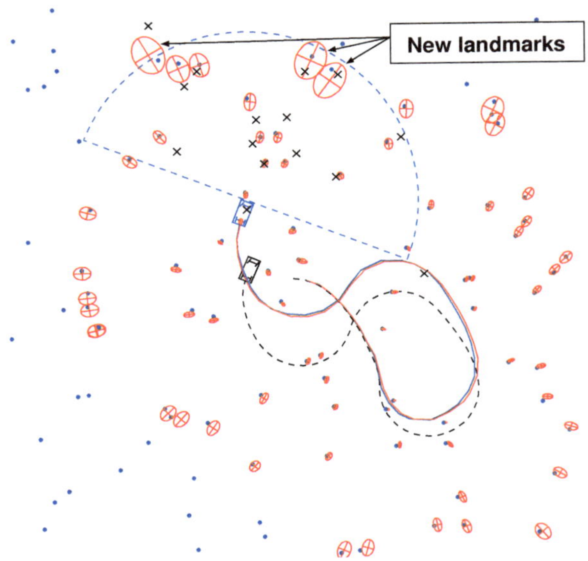

Fig. 3.10. Mutual exclusion helps differentiate between previously observed and new landmarks.

The algorithm shown in Table 3.2 incorporates a single observation for every control. This choice is for notational simplicity only. Multiple readings can be incorporated per time step by processing each observation sequentially. The weight for each particle is equal to the product of the weights due to each observation considered alone. Incorporating multiple observations per time step will increase both the accuracy of data association and the accuracy of the resulting map.

3.6 FastSLAM Extensions

This section describes two extensions to the FastSLAM algorithm. The first, greedy mutual exclusion, improves the accuracy of data association. The second, negative evidence, increases the accuracy of the maps generated by FastSLAM.

3.6.1 Greedy Mutual Exclusion

If multiple observations are incorporated simultaneously, the simplest approach to data association is to consider the identity of each observation independently. However, the data associations of each observation are clearly correlated, as was shown in Section 3.4. The data associations are correlated through error in the robot pose, and they also must all obey a mutual exclusion constraint; more than one observation cannot be associated with the same landmark at the same time. Considering the data associations jointly does address these problems [1, 14, 68], but these techniques are computationally expensive for large numbers of simultaneous observations.

FastSLAM addresses the first problem, motion ambiguity, by sampling over robot poses and data associations. Each set of data association decisions is conditioned on a particular robot path. Thus, the data associations can be chosen independently without fear that pose error will corrupt all of the decisions. Some of the particles will chose the correct data associations, while others draw inconsistent robot poses, pick incorrect data associations, and receive low weights. Picking associations independently per particle still ignores the issue of mutual exclusion, however. Mutual exclusion is particularly useful for deciding when to add new landmarks in noisy environments. Instead of assigning an observation of an unseen landmark to an existing landmark, mutual exclusion will force the creation of a new landmark if both features are observed.

Proper handling of mutual exclusion requires that the data associations of all observations be considered simultaneously. However, mutual exclusion can also be enforced in a greedy fashion. Each observation is processed sequentially and ignores the landmarks associated with previously assigned observations. With a single data association hypothesis, applying mutual exclusion greedily can lead to failures in noisy environments. It can work well in FastSLAM, though, because the motion ambiguity that commonly causes greedy mutual exclusion failures is largely factored out by sampling over the the robot's path. Furthermore, errors due to the greedy nature of the algorithm can also be minimized by processing the observations in different orders for each particle.

3.6.2 Feature Elimination Using Negative Evidence

The point landmark representation of maps is an affirmative representation; in other words, the map describes where landmarks *are* in the world but says nothing about where landmarks *are not*. As such, observations are typically used only as positive evidence of the existence of landmarks in this framework. Observations also can be used, however, to determine whether a landmark in the map actually exists in the world. If the robot predicts that it should see a landmark and then does not, this lack of observation provides evidence that the landmark does not actually exist. These "phantom landmarks" may occur if the robot's sensors generate spurious measurements. The absence of observations of a landmark is sometimes referred to as negative evidence. It can be used to remove false landmarks caused by outlier observations.

Negative evidence can be incorporated into FastSLAM in a simple manner. Let $i_n^{[m]}$ be a binary variable that indicates the existence of feature $\theta_n^{[m]}$. Observing the landmark θ_n provides positive evidence for its existence, whereas not observing the landmark when the landmark falls within the robot's perceptual range provides negative evidence. The resulting posterior probability distribution

$$p(i_n^{[m]} \mid \hat{n}^{t,[m]}, s^{t,[m]}, z^t) \tag{3.54}$$

is estimated using a binary Bayes filter, an algorithm familiar in the literature of occupancy grid maps [63]. FastSLAM represents the posterior over landmark existence in log-odds form:

$$\tau_n^{[m]} = \ln \frac{p(i_n^{[m]} \mid \hat{n}^{t,[m]}, s^{t,[m]}, z^t)}{1 - p(i_n^{[m]} \mid \hat{n}^{t,[m]}, s^{t,[m]}, z^t)} = \sum_t \ln \frac{p(i_n^{[m]} \mid \hat{n}_t^{[m]}, s_t^{[m]}, z_t)}{1 - p(i_n^{[m]} \mid \hat{n}_t^{[m]}, s_t^{[m]}, z_t)} \tag{3.55}$$

The advantage of the log-odds form is that the updates are additive. See [87] for a derivation. Observation of a landmark leads to the addition of a positive value ρ^+ to $\tau_n^{[m]}$, and not observing a landmark that should have been seen results in the addition of a negative value ρ^-. The ρ values are defined as:

$$\rho^+ = \ln \frac{p(i_{\hat{n}_t}^{[m]} \mid \hat{n}_t^{[m]}, s_t^{[m]}, z_t)}{1 - p(i_{\hat{n}_t}^{[m]} \mid \hat{n}_t^{[m]}, s_t^{[m]}, z_t)} \tag{3.56}$$

$$\rho^- = \ln \frac{p(i_{n \neq \hat{n}_t}^{[m]} \mid \hat{n}_t^{[m]}, s_t^{[m]}, z_t)}{1 - p(i_{n \neq n_t}^{[m]} \mid \hat{n}_t^{[m]}, s_t^{[m]}, z_t)} \tag{3.57}$$

The log odds ratio can be implemented easily in real-time. Each landmark filter maintains an estimate of $\tau_n^{[m]}$ as shown above. Fixed values are added or subtracted from τ depending on whether the landmark is observed or missed. If τ falls below a minimum value that corresponds to a minimum probability of existence, then the landmark filter is removed. This mechanism enables FastSLAM particles to free themselves of landmarks caused by spurious measurements.

3.7 Log(N) FastSLAM

The computational complexity of the FastSLAM algorithm presented up to this point requires time $O(M \cdot N)$ where M is the number of particles, and N is the number of landmarks in the map. The linear complexity in M is unavoidable, given that we have to process M particles for every update. This linear complexity in N is due to the importance resampling step in Section 3.3.4. Since the sampling is done with replacement, a single particle in the weighted particle set may be duplicated several times in S_t. The simplest way to implement this is to repeatedly copy the entire particle into the new particle set. Since the length of the particles depends linearly on N, this copying operation is also linear in the size of the map.

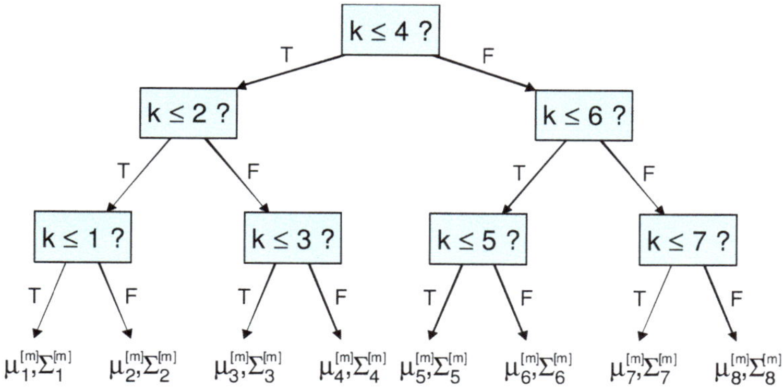

Fig. 3.11. Binary tree of landmark filters.

The wholesale copying of particles from the old set into the new set is an overly conservative approach. The majority of the landmark filters remain unchanged at every time step. Indeed, since the sampling is done with replacement, many of the landmark filters will be completely identical.

These observations suggest that with proper bookkeeping, a more efficient particle representation might allow duplicate landmark filters to be shared between particles, resulting in a more efficient implementation of FastSLAM. This can be done by changing the particle representation from an array of landmark filters to a binary tree. An example landmark tree is shown in Figure 3.11 for a map with eight landmarks. In the figure, the landmarks are organized by an arbitrary landmark number K. In situations in which data association is unknown, the tree could be organized spatially as in a k-d tree.

Note that the landmark parameters μ_n, Σ_n are located at the leaves of the tree. Each non-leaf node in the tree contains pointers to up to two subtrees. Any subtree can be shared between multiple particles' landmark trees. Sharing subtrees makes the update procedure more complicated to implement, but results in a tremendous savings in both memory and computation. Assuming that the tree is balanced, accessing a leaf requires a binary search, which will take $\log(N)$ time, on average.

The $\log(N)$ FastSLAM algorithm can be illustrated by tracing the effect of a control and an observation on the landmark trees. Each new particle in S_t will differ from its generating particle in S_{t-1} in two ways. First, each will posses a different pose estimate from (3.17), and second, the observed feature's Gaussian will be updated as specified in (3.29)-(3.34). All other Gaussians will be equivalent to the generating particle. Thus, when copying the particle to S_t, only a single path from the root of the tree to the updated Gaussian needs to be duplicated. The length of this path is logarithmic in N, on average.

An example is shown in Figure 3.12. Here we assume that $n_t = 3$, that is, only the landmark Gaussian parameters $\mu_3^{[m]}, \Sigma_3^{[m]}$ are updated. Instead

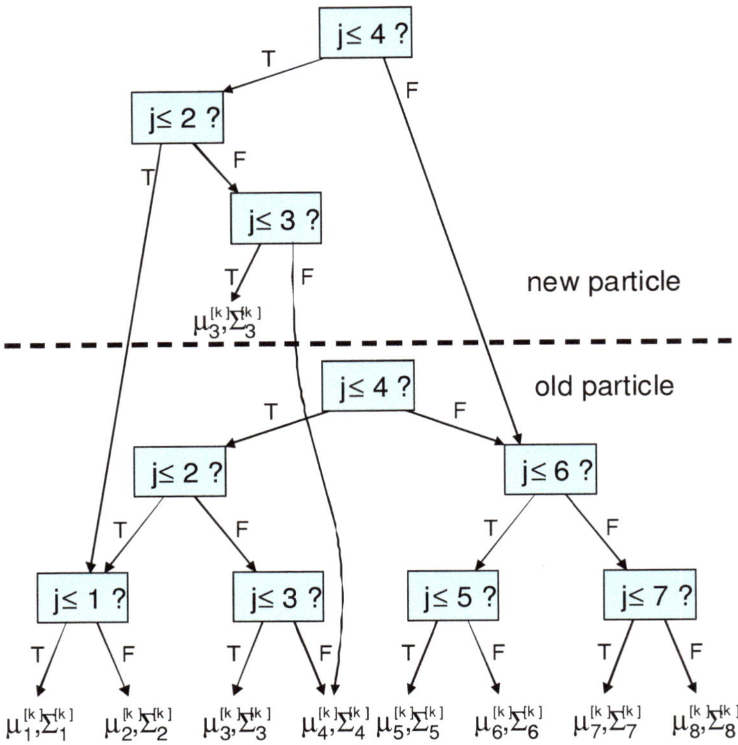

Fig. 3.12. Updating the Landmark Tree.

of duplicating the entire tree, a single path is duplicated, from the root to the third Gaussian. This path is an incomplete tree. The tree is completed by copying the missing pointers from the tree of the generating particle. Thus, branches that leave the modified path will point to the unmodified subtrees of the generating particle. Clearly, generating this modified tree takes time logarithmic in N. Moreover, accessing a Gaussian also takes time logarithmic in N, since the number of steps required to navigate to a leaf of the tree is equivalent to the length of the path. Thus, both generating and accessing a partial tree can be done in time $O(\log N)$. M new particles are generated at every update step, so the resulting FastSLAM algorithm requires time $O(M \log N)$.

3.7.1 Garbage Collection

Organizing particles as binary trees naturally raises the question of garbage collection. Subtrees are constantly being shared and split between particles. When a subtree is no longer referenced as a part of any particle description, the memory allocated to this subtree must be freed. Otherwise, the memory required by FastSLAM will grow without bound.

Whenever landmarks are shared between particles, the shared landmarks always form complete subtrees. In other words, if a particular node of the landmark tree is shared between multiple particles, all of the nodes descendants will also be shared. This greatly simplifies garbage collection in FastSLAM, because the landmark trees can be freed recursively.

Garbage collection in FastSLAM can be implemented using reference counts attached to each node in the landmark tree. Each counter counts the number of times the given node is pointed to by other nodes. A newly created node, for example, receives a reference count of 1. When a new reference is made to a node, the reference count is incremented. When a link is removed, the reference count is decremented. If the reference count reaches zero, the reference counts of the node's children are decreased, and the node's memory is freed. This process is then applied recursively to all children of the node with a zero reference count. This process will require $O(M \log N)$ time on average. Furthermore, it is an optimal deallocation algorithm, in that all unneeded memory is freed immediately when it is no longer referenced.

3.7.2 Unknown Data Association

If the mapping between landmarks and observations is not known, then the landmark trees associated with each particle must be organized spatially, instead of by landmark identity. Kd-trees [62] can guarantee logarithmic time search for high likelihood features, and features in the robot's sensor range. Incremental techniques for constructing balanced kd-trees are described in [52, 76]. The bkd-tree proposed in [76] maintains a sequence of trees of growing complexity. By carefully shifting items across those trees, logarithmic time recall and amortized logarithmic time for landmark insertion should be possible in FastSLAM. Using such a data structure, all necessary operations for FastSLAM with unknown data association could be carried out in logarithmic time on average. $\log(N)$ FastSLAM with unknown data association has not been implemented, so the practical performance of this algorithm is unknown.

3.8 Experimental Results

In the following sections, we present experimental results to validate the performance of the FastSLAM algorithm. We will start by demonstrating the performance of FastSLAM on data collected by a real robot, and then go on to evaluate specific aspects of the algorithm on simulated data. The performance of FastSLAM will be compared against that of the EKF.

Fig. 3.13. University of Sydney High Speed Vehicle (HSV) in Victoria Park.

3.8.1 Victoria Park

The FastSLAM algorithm was tested on a benchmark SLAM data set from the University of Sydney [37]. An instrumented vehicle, shown in Figure 3.13, equipped with a laser range finder was repeatedly driven through Victoria Park, in Sydney, Australia. Victoria Park is an ideal setting for testing SLAM algorithms because the park's trees are distinctive features in the robot's laser scans. Encoders measured the vehicle's velocity and steering angle. Range and bearing measurements to nearby trees were extracted from the laser data using a local minima detector. The vehicle was driven around for approximately 30 minutes, covering a distance of over 4 km. The vehicle is also equipped with GPS in order to capture ground truth data. Due to occlusion by foliage and buildings, ground truth data is only available for part of the overall traverse. While ground truth is available for the robot's path, no ground truth data is available for the locations of the landmarks.

Since the robot is driving over uneven terrain, the measured controls are fairly noisy. Figure 3.14(a) shows the path of the robot obtained by integrating the estimated controls. After 30 minutes of driving, the estimated position of the robot is well over 100 meters away from its true position measured by GPS. The laser data, on the other hand, is a very accurate measure of range and bearing. However, not all objects in the robot's field of view are trees, or even static objects. As a result, the feature detector produced relatively accurate observations of trees, but also generated frequent outliers.

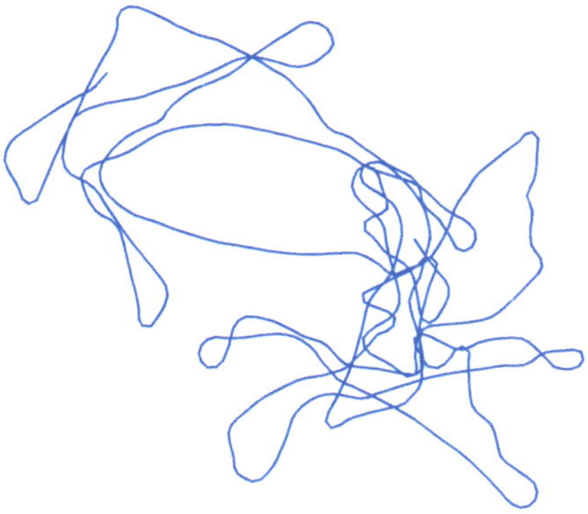

(a) Vehicle path predicted by the odometry

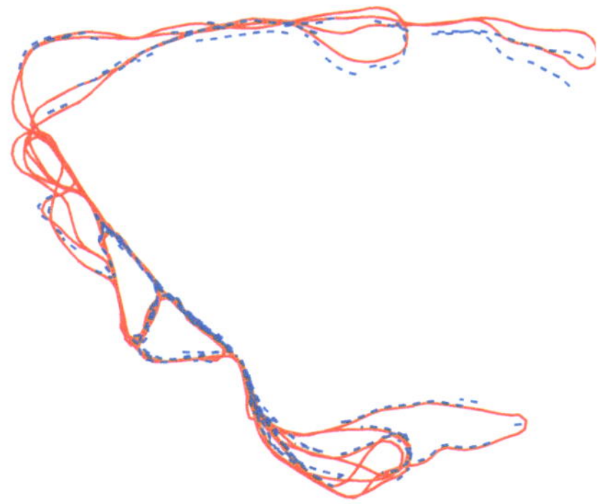

(b) True path (dashed line) and FastSLAM path (solid line)

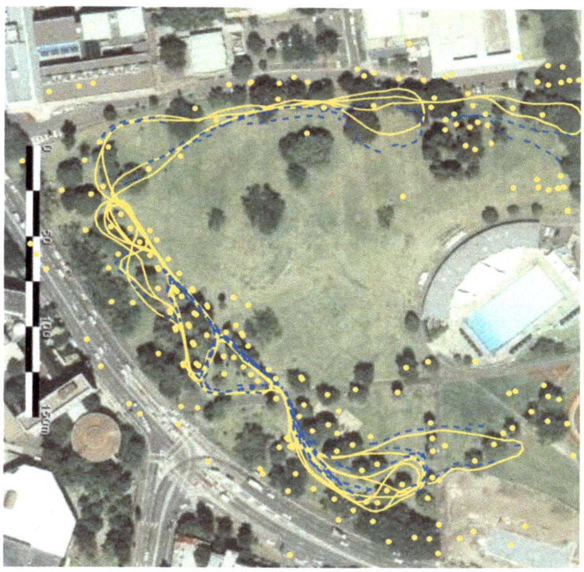

(c) Victoria Park results overlaid on aerial imagery with the GPS path in blue (dashed), average FastSLAM path in yellow (solid), and estimated landmarks as yellow circles

Fig. 3.14. Results of FastSLAM on Victoria Park data set.

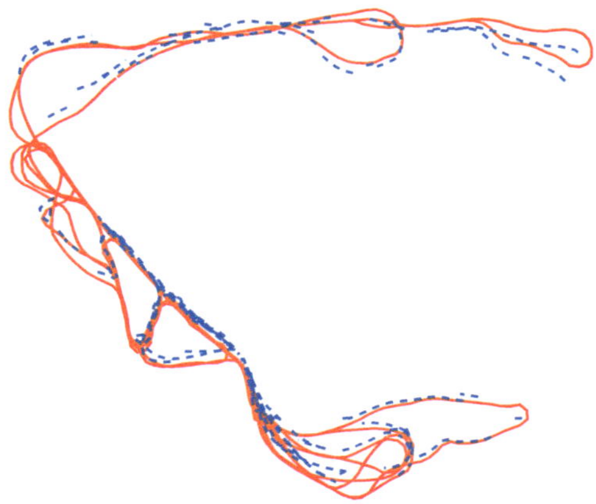

Fig. 3.15. Victoria Park Map created without odometry information.

Data association for this experiment was done using per-particle ML data association. Since the accuracy of the observations is high relative to the average density of landmarks, data association in the Victoria Park data set is a relatively straightforward problem. In a later experiment, more difficult data association problems will be simulated by adding extra control noise.

The output of FastSLAM is shown in Figure 3.14(b) and (c). The GPS path is shown as a dashed line, and the output of FastSLAM is shown as a solid line. The RMS error of the resulting path is just over 4 meters over the 4 km traverse. This experiment was run with 100 particles.

Performance Without Odometry

FastSLAM was also run on the Victoria Park data set without using the odometry data. The pose of the robot in each particle was supplemented with translational velocity v_t and rotational velocity w_t.

$$S_t^{[m]} = \langle s_{x,t}, s_{y,t}, s_{\theta,t}, s_{v,t}, s_{w,t}, N_t, \mu_{1,t}, \Sigma_{1,t}, \ldots, \mu_{N_t,t}, \Sigma_{N_t,t} \rangle \tag{3.58}$$

A Brownian motion model was used to predict the pose of the robot at time $t+1$ given the pose at time t. This model assumes that the velocity at time $t+1$ is equal to the velocity at time t plus some random perturbation.

$$v_t = v_{t-1} + \mathcal{N}(v; 0, \alpha_1^2) \tag{3.59}$$

$$w_t = w_{t-1} + \mathcal{N}(w; 0, \alpha_2^2) \tag{3.60}$$

After drawing a perturbed velocity, the robot's position is updated accordingly.

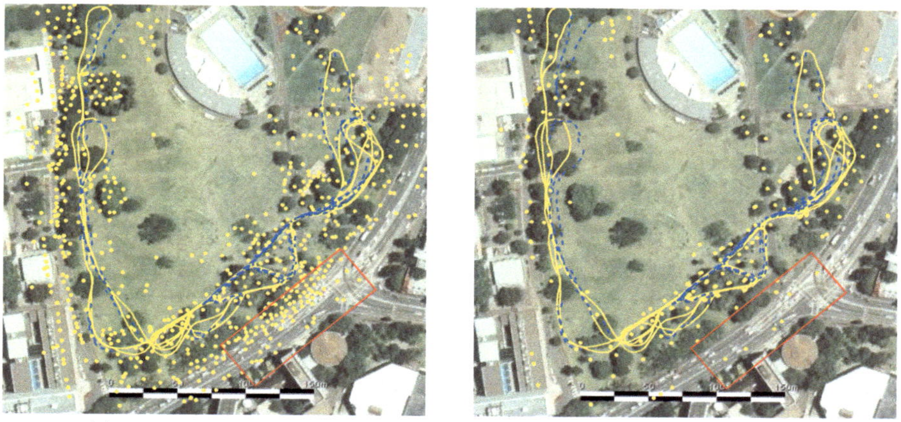

(a) Standard FastSLAM (b) Using negative information

Fig. 3.16. Maps created with and without negative information.

$$x_t = x_{t-1} + v_t \cos(\theta_{t-1})\Delta t \tag{3.61}$$

$$y_t = y_{t-1} + v_t \sin(\theta_{t-1})\Delta t \tag{3.62}$$

$$\theta_t = \theta_{t-1} + w_t \Delta t \tag{3.63}$$

The specific values of α_1 and α_2 depend on the maximum translational and rotational accelerations that the robot is capable of executing. The map created without using the odometry is shown in Figure 3.15. The average error of the map is equivalent to the results obtained with odometry.

Negative Information

In the Victoria Park data set, observations corresponding to non-point objects or non-static objects result in a large number of spurious landmarks being added to every FastSLAM particle. When negative information is used to estimate the existence of each landmark, as described in Section 3.6.2, many of these spurious landmarks can be removed. In the case of Victoria Park, use of negative information results in 44% percent fewer landmarks in the resulting map. While the "correct" number of landmarks is not available, visual inspection of the maps suggests that many of the spurious features have been eliminated. Figure 3.16 shows the Victoria Park map built with and without considering negative evidence. The number of landmarks in areas that should be free of landmarks (the roadway, highlighted with a box in the figure) has been significantly reduced.

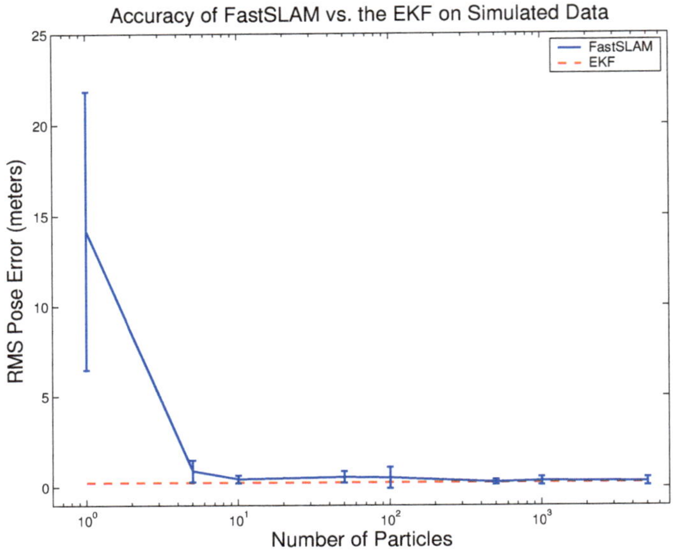

Fig. 3.17. A comparison of the accuracy of FastSLAM and the EKF on simulated data.

3.8.2 Comparison of FastSLAM and the EKF

Accuracy

The accuracy of FastSLAM was compared with that of the EKF on a simulated data set with 100 landmarks. The RMS robot pose error was computed for FastSLAM for various numbers of particles from 1 to 5000. Each experiment was run 10 times. The results are shown in Figure 3.17. The error of the EKF is shown as a dashed horizontal line.

In this experiment, the accuracy of FastSLAM approaches the accuracy of the EKF as the number of particles is increased. Most notably, the error of FastSLAM becomes statistically indistinguishable from that of the EKF past approximately 10 particles. This is interesting because FastSLAM with 10 particles and 100 landmarks requires an order of magnitude fewer parameters than the EKF in order to achieve this level of accuracy. Clearly, the specific value of this threshold of performance will depend on both the parameters of the motion and measurement model and the robot's control policy. However, this experiment suggests that in normal circumstances, a relatively small number of particles may suffice to achieve high estimation accuracy.

Scaling Performance

The scaling performance of FastSLAM was also evaluated on simulated data. Simulated maps of constant landmark density were created with varying num-

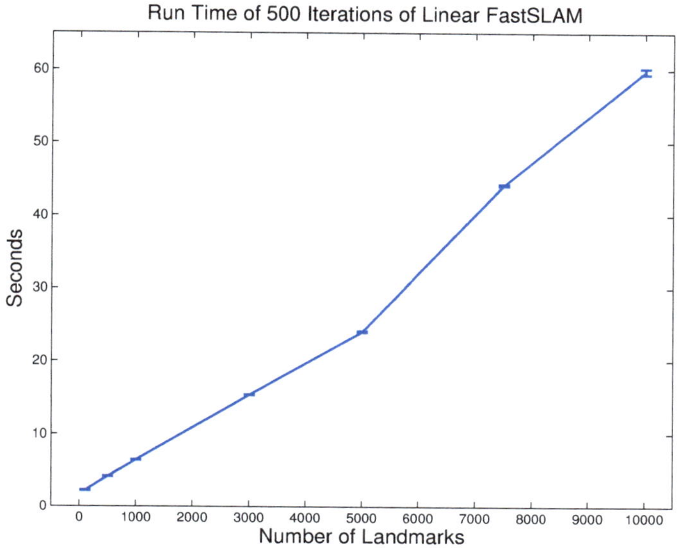

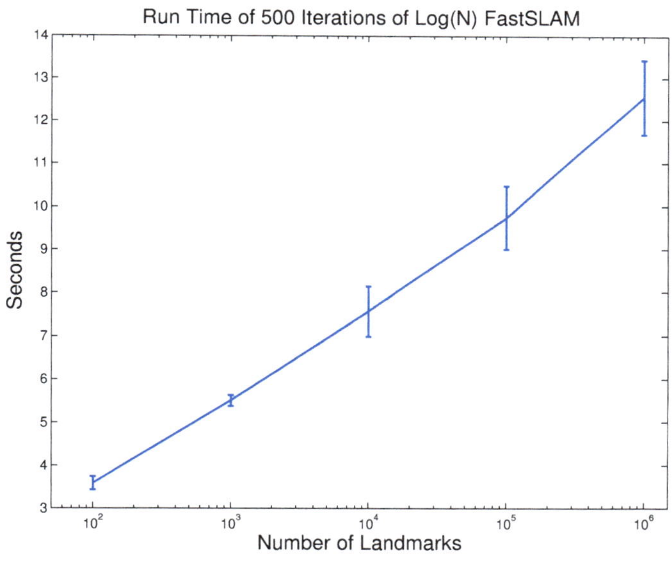

Fig. 3.18. Timing results for FastSLAM in simulated environments.

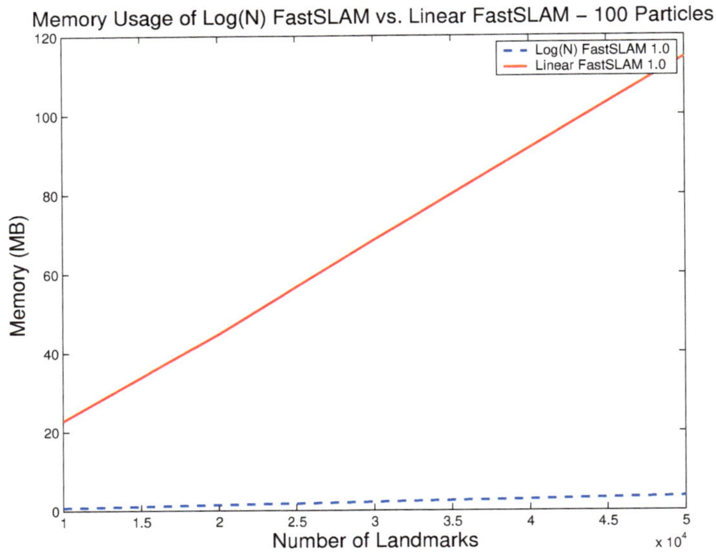

Fig. 3.19. Memory requirements for linear and log(N) version of FastSLAM in simulated environments.

bers of landmarks. Constant landmark density ensures that the simulated robot observed a constant number of landmarks on average across all trials. The performance of the linear time and logarithmic time versions of the Fast-SLAM algorithm were compared. The linear time algorithm was tested with up to 10,000 landmarks, and the logarithmic time algorithm was tested with up to 1,000,000 landmarks. The time required to compute 500 sensor updates with all landmarks incorporated into the map was evaluated over 10 different runs. All experiments were done with 100 particles.

The results of the experiment are shown in Figure 3.18. The performance of the log(N) algorithm is plotted on a logarithmic scale. The results validate the scaling performance of the tree-based algorithm, and demonstrate the substantial performance increase enabled by sharing landmark trees across particles.

Sharing subtrees is not only computationally efficient; it also decreases the overall memory required by the algorithm. The memory required by both versions of the FastSLAM algorithm scales linearly with the number of landmarks. Overall, the FastSLAM algorithm must maintain $M \cdot N$ landmark filters. With 100 particles and 1,000,000 landmarks, this can add up to a substantial amount of memory (hundreds of megabytes) just to represent the map. In very large maps, landmarks that have not been visited for a long period of time will be shared in subtrees between all of the particles of the log(N) algorithm. If only a fraction of the total landmarks are observed at every time step, this memory sharing may result in a significant savings in memory consumption. A plot of the memory consumed by the linear and logarithmic FastSLAM algorithms for varying numbers of landmarks is shown in Figure 3.19. In this experiment, the tree-based representation resulted in over an order-of- magnitude decrease in memory consumption over the basic FastSLAM algorithm.

3.8.3 Ambiguous Data Association

The performance of FastSLAM given unknown data association was evaluated against that of the Extended Kalman Filter using the Victoria Park data set. Under normal conditions, the levels of odometric and measurement noise present in the Victoria Park data set do not cause a significant data association problem. The error of the vehicle's laser is quite low compared to the average distance between trees in the park. In order to test performance given data association ambiguity, additional odometric noise was added to the robot controls. Additional control noise results in high motion ambiguity in the data associations of new observations.

Prototypical outputs of the EKF and FastSLAM given low and high levels of odometric noise are shown in Figure 3.20. While both algorithms generate accurate maps when control noise is low, the EKF fails catastrophically with high error. The map generated by FastSLAM under high odometric error is not degraded in quality. The RMS error of the robot position was computed for FastSLAM and the EKF over 20 different runs with four different levels of odometric noise. The results are shown in Figure 3.21. As control noise increases, there is no measurable increase in the RMS error of FastSLAM, while the error of the robot path emitted by the EKF goes up substantially. More telling is the variance in the error of the EKF maps across multiple runs, indicated by the confidence bars. This suggests that for high levels of control noise, the EKF is diverging.

3.8.4 Sample Impoverishment

FastSLAM, like all particle filters, works best if the proposal distribution and the posterior distribution are well matched. If the robot's motion is very noisy and the robot's sensor is very accurate, many particles will be thrown away in the resampling process. In these situations, the performance of FastSLAM will degrade. This effect can be demonstrated by running FastSLAM on simulated data with varying levels of sensor noise. The results of this experiment are shown in Figure 3.22. FastSLAM was run with 100 particles for various levels of sensor noise. Each experiment was run 10 times. Clearly, as the measurement error becomes very low, FastSLAM begins to diverge. It is important to note that these values of measurement error are very small (less than 1% of the average range value). This problem will be addressed in the following chapter.

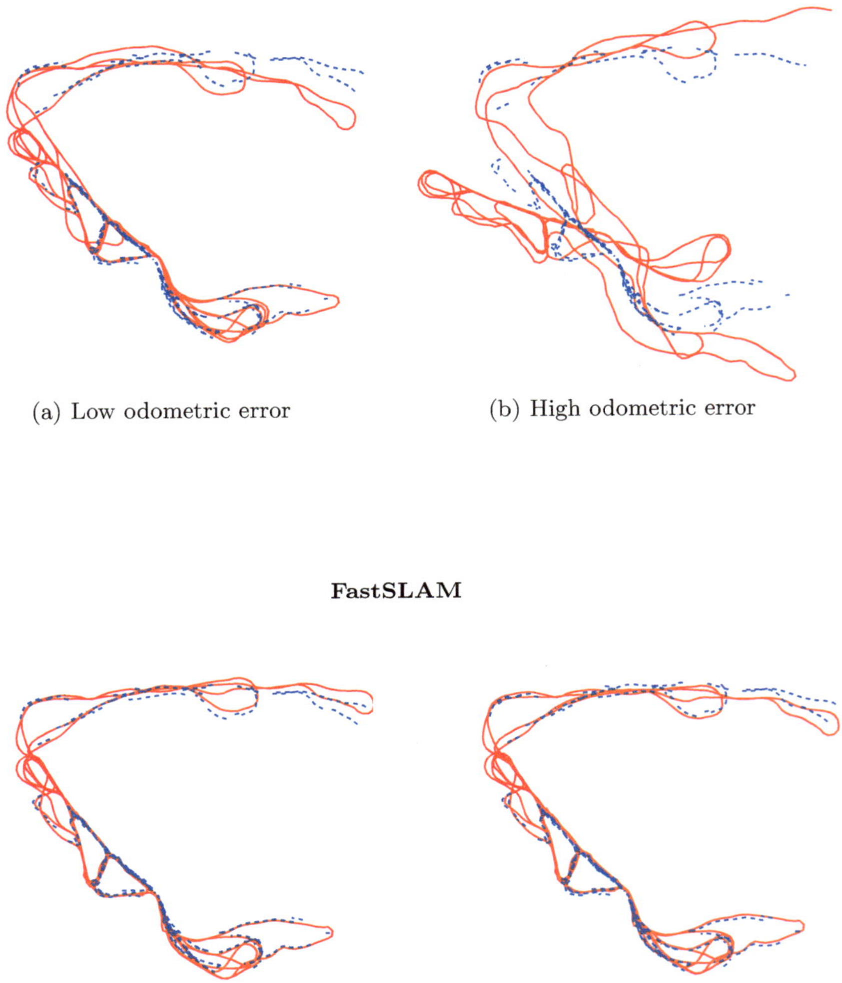

Extended Kalman Filter

(a) Low odometric error

(b) High odometric error

FastSLAM

(c) Low odometric error

(d) High odometric error

Fig. 3.20. Performance of EKF and FastSLAM on the Victoria Park data set with varying levels of odometric noise.

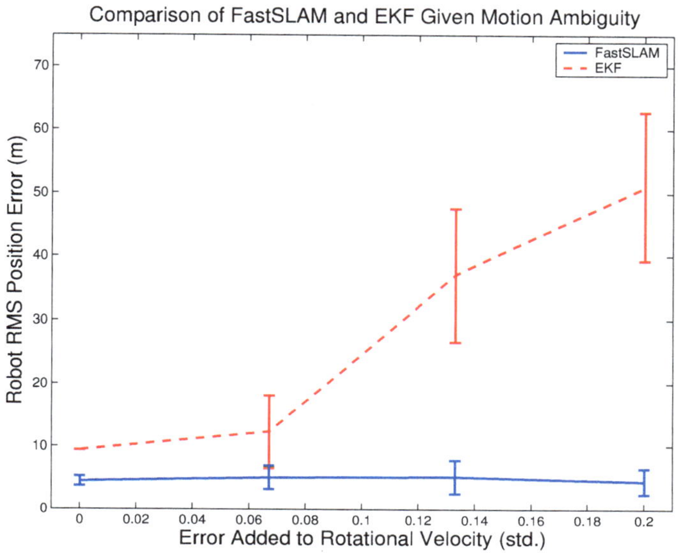

Fig. 3.21. Position error of vehicle under various levels of odometric noise.

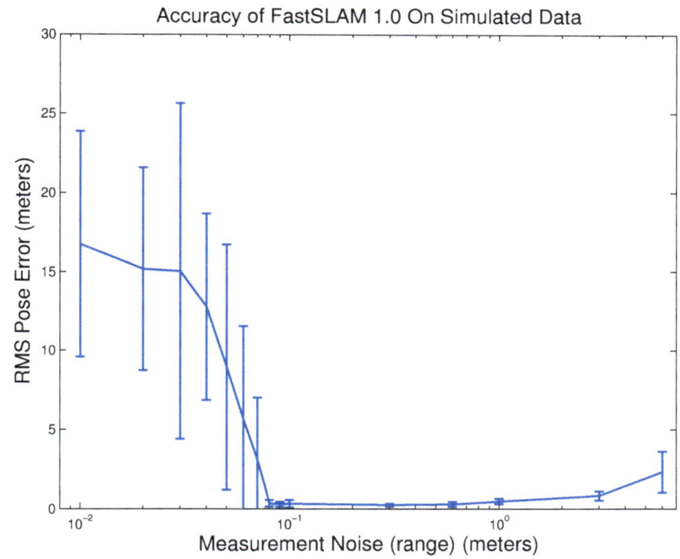

Fig. 3.22. FastSLAM 1.0 diverges if the robot's sensor is too accurate relative to the robot's motion error.

3.9 Summary

This chapter presented FastSLAM, a SLAM algorithm based on particle filtering. FastSLAM samples over robot pose and data associations, and computes the positions of the landmarks conditioned on each particle. This factorization coupled with a binary tree representation of the landmarks, enables sensor updates to be performed in logarithmic time. Per-particle data association factors robot pose uncertainty out of the data association problem. This, in turn, results in better discrimination between landmarks in noisy environments, even with relatively simple data association heuristics.

Experimental results verify the computational savings of FastSLAM. The $\log(N)$ version of the algorithm has been tested in environments with millions of landmarks. Given known data association, the accuracy of FastSLAM approached the accuracy of the EKF given a sufficient number of particles. If the data associations were unknown, then FastSLAM significantly outperformed the EKF on both real and simulated data sets. The accuracy of the basic FastSLAM algorithm does degrade if the proposal distribution and the posterior are mismatched. The following chapter will describe an extension of the FastSLAM that addresses this problem.

4

FastSLAM 2.0

This chapter presents and improved version of the FastSLAM algorithm, called FastSLAM 2.0.

Sampling over robot paths leads to efficient scaling and robust data association, however it also has its drawbacks. FastSLAM, and particle filters in general, have some unusual properties. For example, the performance of the algorithm will eventually degrade if the robot's sensor is *too* accurate. This problem occurs when the proposal distribution is poorly matched with the posterior. In FastSLAM, this happens when the motion of the robot is noisy relative to the observations. This chapter will describe a modified version of FastSLAM, called FastSLAM 2.0 [60], which attempts to solve this problem. FastSLAM 2.0 incorporates the current observation into the proposal distribution, not just the importance weights, in order to better match the posterior. The resulting algorithm is superior to the original FastSLAM algorithm in nearly all respects. FastSLAM 2.0's main disadvantage is that it is more difficult to implement that FastSLAM 1.0; it is also more mathematically involved.

4.1 Sample Impoverishment

The two key steps of particle filtering, sampling from the proposal distribution and importance resampling, can be thought of as a process of incrementally building and pruning a tree of representative trajectories of the system being filtered. Drawing from the proposal distribution generates new trajectories, and resampling throws away very improbable trajectories in order to maintain a constant number of particles. Clearly, a particle filter with many distinct trajectories is a more diverse sampling of the posterior than a particle filter with many duplicate trajectories. In general, more diversity in the sample set will lead to better estimation accuracy.

The resampling process, while an essential part of the particle filter, decreases diversity by throwing away some particles and duplicating others mul-

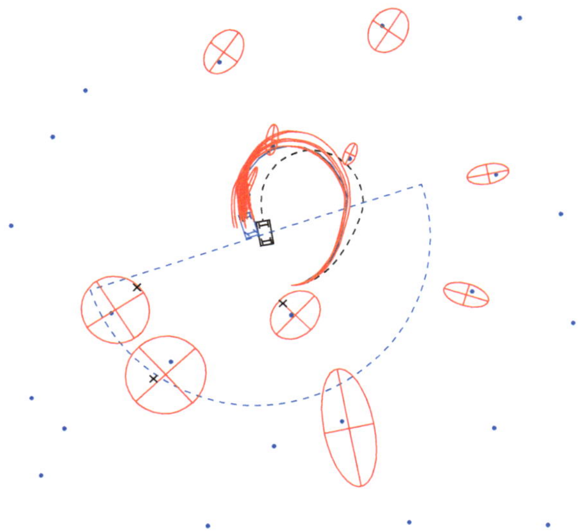

Fig. 4.1. Particle history in FastSLAM.

tiple times. As many controls and observations are incorporated into the filter, all of the particles will inevitably share some common history. In other words, if the trajectory of every particle is traced back in time, at some point all of the particles will share a single common ancestor. The number of generations that the particles can be traced back through time to a common estimator is an important indicator of the performance of the filter.

In the case of FastSLAM, the histories of the particles have a particularly intuitive form; each is a hypothesized path of the robot. Conditioned on each robot path, the landmark positions are independent. The correlations between the landmarks are represented in the collection of robot particles. The more diverse the particle set, the more accurately FastSLAM can revise the path of the robot (and thus the landmark positions) given a new observation. The path distance back to the common trajectory can be thought of as the distance beyond which the filter cannot go back and re-weight past hypotheses. This distance is crucial to the performance of FastSLAM, as it defines how large a loop can effectively be closed. The histories of particles is shown for a simulated FastSLAM run in Figure 4.1.

The amount of diversity in the FastSLAM sample set is governed by the balance between the proposal and pruning processes. The more hypotheses that are pruned during every update, the lower the diversity will be. The mismatch between the proposal and posterior distribution in FastSLAM is most pronounced when the noise of robot motion is much higher than the noise of the robot's sensors. The motion model spreads the particles out over

Fig. 4.2. Mismatch between proposal and posterior distributions.

a large space, but only a small fraction of the particles receive non-negligible weights when the observation is incorporated. Thus, only a small fraction of the samples survive into the next generation. This problem is compounded further if the robot incorporates a large number of observations per control.

Figure 4.2 shows an example of a mismatched proposal and posterior distribution. The particles in Figure 4.2(a) are drawn from the motion model, which is substantially noisy. A subsequent observation constrains the robot's true position to lie within the overlaid ellipse. All of the samples outside of the ellipse, grayed out in Figure 4.2(a), will receive negligible probability, while the samples inside the ellipse will be duplicated multiple times by the resampling process. As the observations become more accurate, fewer unique samples will survive each update. At some point, sufficiently accurate observations will cause the particle filter to diverge. This effect was demonstrated in Figure 3.22.

4.2 FastSLAM 2.0

This chapter describes a modified version of the basic FastSLAM algorithm designed to be more efficient with its samples. Conceptually this modification is simple: when sampling a new robot pose, the new proposal distribution will rely not only on the current control (as is the case in FastSLAM 1.0), but also on the most recent sensor measurement. To obtain a suitable proposal distribution, the FastSLAM 2.0 algorithm linearizes the motion model, in the same manner as EKF-based SLAM solutions. As a result, the modified proposal distribution can be computed in closed form.

This extension parallels prior work by Doucet et al., who proposed a similar modification for general particle filters [21], and Markov Chain Monte Carlo techniques for neural networks [11]. It is also similar to the arc reversal technique proposed for particle filters applied to Bayes Networks [72], and work by

Van de Merwe et al., which uses a unscented filtering step [92] for generating proposal distributions that incorporate the most recent measurement.

While this modification is conceptually simple, it has important ramifications. First, it leads to a proof of convergence for the FastSLAM 2.0 algorithm in linear-Gaussian worlds. The resulting one-particle algorithm requires constant time to incorporate observations. The best previous SLAM algorithm for which convergence was shown requires quadratic update time. Furthermore, the new algorithm, even with a single particle, yields significantly more accurate results than FastSLAM 1.0 on the Victoria Park dataset. These findings are significant, as many mobile robot systems are plagued by control noise, but possess relatively accurate sensors.

Incorporating the modified proposal distribution requires two primary changes to the FastSLAM algorithm. First, a procedure for drawing samples from the new proposal must be developed. Second, the formula for the weights of the particles must be updated to reflect the change in proposal distribution.

4.2.1 The New Proposal Distribution

In FastSLAM 1.0, the proposal distribution produces new robot poses that agree with the current control u_t. The FastSLAM 2.0 proposal distribution is designed to generate new robot poses that also agree with the current observation z_t. If the robot's motion model is non-linear, observations cannot be incorporated into the proposal in closed form. FastSLAM 2.0 solves this problem by linearizing the motion model. Drawing a sample from the new proposal distribution given each particle in S_{t-1} is a three step process:

1. Project the previous pose of the robot $s_{t-1}^{[m]}$ forward according to the linear-Gaussian motion model. The uncertainty of the resulting pose $\hat{s}_t$ will be a Gaussian.
2. Consider this Gaussian to be the prior distribution of an EKF estimating the pose of the robot. Incorporate the observation z_t into this EKF using the standard EKF update equations.
3. Draw a sample $s_t^{[m]}$ from the resulting posterior distribution.

The Gaussian distribution resulting from the one-step EKF update will have a smaller uncertainty than both the motion model Gaussian and the observation Gaussian projected into robot pose coordinates. This update can be thought of as a process for creating "virtual controls" by matching observations to the map. If the observations are very accurate, the virtual controls will often be more accurate than the actual controls. In these situations, the FastSLAM 2.0 proposal distribution will significantly outperform the original proposal distribution.

Derivation of the FastSLAM Proposal Distribution

Much like the derivation of the derivation of the original FastSLAM algorithm, we will represent the standard motion and measurement models as nonlinear functions with independent Gaussian noise:

$$p(z_t \mid s_t, \Theta, n_t) = g(s_t, \theta_{n_t}) + \varepsilon_t \tag{4.1}$$
$$p(s_t \mid u_t, s_{t-1}) = h(s_{t-1}, u_t) + \delta_t \tag{4.2}$$

Here g and h are nonlinear functions, and ε_t and δ_t are Gaussian noise variables with covariance R_t and P_t, respectively.

Instead of drawing a new pose s_t from the standard motion model $p(s_t \mid u_t, s_{t-1})$, FastSLAM 2.0 will draw a new pose from a motion model that includes the most recent observation z_t.

$$s_t^{[m]} \sim p(s_t \mid s^{t-1,[m]}, u^t, z^t, n^t) \tag{4.3}$$

Here $s^{t-1,[m]}$ is the path up to time $t-1$ attached to the m-th particle. The sampling distribution (4.3) explicitly incorporates the most recent sensor measurement z_t, its data association n_t, and the most recent control u_t, which together represent the new information available at time t.

The mechanism for sampling from (4.3) requires further analysis. First, the new motion model can be rewritten in terms of known distributions, including the standard motion and measurement models, and the Gaussian feature estimates. This derivation is analogous to the derivation of the basic filter equation. First, the proposal distribution is expanded using Bayes Rule.

$$p(s_t \mid s^{t-1,[m]}, u^t, z^t, n^t)$$
$$\overset{Bayes}{\propto} \eta\, p(z_t \mid s_t, s^{t-1,[m]}, u^t, z^{t-1}, n^t)\, p(s_t \mid s^{t-1,[m]}, u^t, z^{t-1}, n^t) \tag{4.4}$$

The robot pose s_t in the second term depends only on the previous pose s_{t-1} and the current control u_t. The rest of the conditioning variables can be dropped, leaving the standard motion model.

$$\overset{Markov}{=} \eta\, p(z_t \mid s_t, s^{t-1,[m]}, u^t, z^{t-1}, n^t)\, p(s_t \mid s_{t-1}^{[m]}, u_t) \tag{4.5}$$

The Theorem of Total Probability is used to condition the first term of the product on the currently observed landmark θ_{n_t}, as follows:

$$= \eta \int p(z_t \mid \theta_{n_t}, s_t, s^{t-1,[m]}, u^t, z^{t-1}, n^t)\, p(\theta_{n_t} \mid s_t, s^{t-1,[m]}, u^t, z^{t-1}, n^t)$$
$$d\theta_{n_t} p(s_t \mid s_{t-1}^{[m]}, u_t) \tag{4.6}$$

The first term of the integrand is simply the measurement model $p(z_t \mid s_t, \theta_{n_t}, n_t)$. The second term can also be simplified because s_t, n_t and u_t do not provide any information about θ_{n_t} without z_t.

$$\stackrel{Markov}{=} \eta \int \underbrace{p(z_t \mid \theta_{n_t}, s_t, n_t)}_{\sim\mathcal{N}(z_t; g(s_t, \theta_{n_t}), R_t)} \underbrace{p(\theta_{n_t} \mid s^{t-1,[m]}, z^{t-1}, u^{t-1}, n^{t-1})}_{\sim\mathcal{N}(\theta_{n_t}; \mu^{[m]}_{n_t,t-1}, \Sigma^{[m]}_{n_t,t-1})} d\theta_{n_t}$$

$$\underbrace{p(s_t \mid s^{[m]}_{t-1}, u_t)}_{\sim\mathcal{N}(s_t; h(s^{[m]}_{t-1}, u_t), P_t)} \quad (4.7)$$

The expression (4.7) shows that the sampling distribution is the convolution of two Gaussians, multiplied by a third. Unfortunately, in the general case this distribution possesses no closed form from which we can easily draw samples. The culprit is the function g; if it were linear, this probability would be Gaussian, a fact that shall become obvious below.

As a result of this observation, the function g will be replaced by a linear approximation. This approximation is obtained through a first order Taylor expansion. The Taylor expansion of g is the following linear function:

$$\hat{s}_t = h(s^{[m]}_{t-1}, u_t) \quad (4.8)$$

$$\hat{z}_t = g(\hat{s}_t, \mu^{[m]}_{n_t,t-1}) \quad (4.9)$$

$$G_\theta = \nabla_{\theta_{n_t}} g(s_t, \theta_{n_t})\big|_{s_t=\hat{s}_t, \theta_{n_t}=\mu^{[m]}_{n_t,t-1}} \quad (4.10)$$

$$G_s = \nabla_{s_t} g(s_t, \theta_{n_t})\big|_{s_t=\hat{s}_t, \theta_{n_t}=\mu^{[m]}_{n_t,t-1}} \quad (4.11)$$

$$g(s_t, \theta_{n_t}) \approx \hat{z}_t + G_\theta(\theta_{n_t} - \mu^{[m]}_{n_t,t-1}) + G_s(s_t - \hat{s}_t) \quad (4.12)$$

Here $\hat{s}_t$ can be thought of as the predicted pose of the robot at time t, and $\hat{z}_t$ as the predicted observation, for the given particle. The matrices G_θ and G_s are the Jacobians of g; that is, they are the derivatives of g with respect to θ_{n_t} and s_t, respectively, evaluated at the expected values of their arguments. Given this linear approximation, we can now determine an approximate form for the proposal distribution. The convolution theorem provides us with a closed form for the integral term in (4.7):

$$\mathcal{N}(z_t; \hat{z}_t + G_s s_t - G_s \hat{s}_t, \underbrace{R_t + G_\theta \Sigma^{[m]}_{n_t,t-1} G_\theta^T}_{Z_t}) \quad (4.13)$$

For brevity, we will write the covariance of this Gaussian as Z_t. Note that this is the same Z_t from the original FastSLAM algorithm, the innovation covariance matrix. The proposal distribution is given by the product of this new Gaussian with the rightmost term in (4.7), the Gaussian $\mathcal{N}(s_t; \hat{s}^{[m]}_t, P_t)$. Expanding the form of the Gaussian explicitly, the product can be written as:

$$p(s_t \mid s^{t-1,[m]}, z^t, u^t, n^t) = \xi \exp\{-y_t\} \quad (4.14)$$

where the exponent is:

$$y_t = \frac{1}{2}[(z - \hat{z}_t - G_s s_t + G_s \hat{s}_t)^T Z_t^{-1}(z - \hat{z}_t - G_s s_t + G_s \hat{s}_t) + (s_t - \hat{s}_t)^T P_t^{-1}(s_t - \hat{s}_t)] \quad (4.15)$$

The expression for the exponent is clearly quadratic in the target variable s_t, therefore the distribution (4.14) is also Gaussian. The mean and covariance of (4.14) are given by the minimum of (4.15) and its curvature. These are identified by calculating the first and second derivatives of y_t with respect to s_t:

$$\frac{\partial y_t}{\partial s_t} = -G^T Z_t^{-1}(z_t - \hat{z}_t - G_s s_t + G_s \hat{s}_t) + P_t^{-1}(s_t - \hat{s}_t) \tag{4.16}$$

$$= (G_s^T Z_t^{-1} G_s + P_t^{-1}) s_t - G_s^T Z_t^{-1}(z_t - \hat{z}_t + G_s \hat{s}_t) - P_t^{-1} \hat{s}_t \tag{4.17}$$

$$\frac{\partial^2 y_t}{\partial s_t^2} = G_s^T Z_t^{-1} G_s + P_t^{-1} \tag{4.18}$$

The covariance $\Sigma_{s_t}^{[m]}$ of the sampling distribution is now obtained by taking the inverse of the second derivative (4.18).

$$\Sigma_{s_t}^{[m]} = \left[G_s^T Z_t^{-1} G_s + P_t^{-1}\right]^{-1} \tag{4.19}$$

The mean $\mu_{s_t}^{[m]}$ of the sample distribution is obtained by setting the first derivative (4.17) to zero, which gives:

$$\begin{aligned}
\mu_{s_t}^{[m]} &= \Sigma_{s_t}^{[m]} \left[G_s^T Z_t^{-1}(z_t - \hat{z}_t + G_s \hat{s}_t^{[m]}) + P_t^{-1} \hat{s}_t\right] \\
&= \Sigma_{s_t}^{[m]} G_s^T Z_t^{-1}(z_t - \hat{z}_t) + \Sigma_{s_t}^{[m]} \left[G_s^T Z_t^{-1} G_s + P_t^{-1}\right] \hat{s}_t^{[m]} \\
&= \Sigma_{s_t}^{[m]} G_s^T Z_t^{-1}(z_t - \hat{z}_t) + \hat{s}_t^{[m]}
\end{aligned} \tag{4.20}$$

(4.19) and (4.20) parameterize FastSLAM 2.0's Gaussian approximation to the new sampling distribution $p(s_t \mid s^{t-1,[m]}, z^t, u^t, n^t)$. This Gaussian is constructed for each particle in S_{t-1}, and a new sample is drawn and placed in the temporary particle set.

4.2.2 Calculating the Importance Weights

Since we are using a different proposal distribution than the original FastSLAM algorithm, the importance weights must also be updated to reflect this change. The importance weights are defined as the ratio of the target distribution over the proposal distribution. Under the asymptotically correct assumption that the paths in $s^{t-1,[m]}$ were generated according to the target distribution one time step earlier, $p(s^{t-1,[m]} \mid z^{t-1}, u^{t-1}, n^{t-1})$,we note that the proposal distribution is given by the product:

$$p(s^{t-1,[m]} \mid z^{t-1}, u^{t-1}, n^{t-1})\, p(s_t^{[m]} \mid s^{t-1,[m]}, z^t, u^t, n^t) \tag{4.21}$$

So the importance weight $w_t^{[m]}$ is equal to:

$$w_t^{[m]} = \frac{\text{target distribution}}{\text{proposal distribution}} \tag{4.22}$$

$$= \frac{p(s^{t,[m]} \mid z^t, u^t, n^t)}{p(s^{t-1,[m]} \mid z^{t-1}, u^{t-1}, n^{t-1})\, p(s_t^{[m]} \mid s^{t-1,[m]}, z^t, u^t, n^t)} \tag{4.23}$$

The numerator can be expanded using the definition of conditional probability:

$$= \frac{p(s_t^{[m]} \mid s^{t-1,[m]}, z^t, u^t, n^t)\, p(s^{t-1,[m]} \mid z^t, u^t, n^t)}{p(s^{t-1,[m]} \mid z^{t-1}, u^{t-1}, n^{t-1})\, p(s_t^{[m]} \mid s^{t-1,[m]}, z^t, u^t, n^t)} \tag{4.24}$$

$$= \frac{p(s^{t-1,[m]} \mid z^t, u^t, n^t)}{p(s^{t-1,[m]} \mid z^{t-1}, u^{t-1}, n^{t-1})} \tag{4.25}$$

Next, the numerator can be expanded using Bayes Rule.

$$\stackrel{Bayes}{=} \eta\, \frac{p(z_t \mid s^{t-1,[m]}, z^{t-1}, u^t, n^t)\, p(s^{t-1,[m]} \mid z^{t-1}, u^t, n^t)}{p(s^{t-1,[m]} \mid z^{t-1}, u^{t-1}, n^{t-1})} \tag{4.26}$$

Here u_t and n_t can be dropped from the second term in the numerator by the Markov property.

$$\stackrel{Markov}{=} \eta\, \frac{p(z_t \mid s^{t-1,[m]}, z^{t-1}, u^t, n^t)\, p(s^{t-1,[m]} \mid z^{t-1}, u^{t-1}, n^{t-1})}{p(s^{t-1,[m]} \mid z^{t-1}, u^{t-1}, n^{t-1})} \tag{4.27}$$

$$= \eta\, p(z_t \mid s^{t-1,[m]}, z^{t-1}, u^t, n^t) \tag{4.28}$$

This equation is similar, but not identical, to the formula for the importance weights in FastSLAM 1.0 (3.39). Next, we apply the Theorem of Total Probability twice in order to condition this expression on s_t and θ_{n_t}.

$$w_t^{[m]} = \eta \int p(z_t \mid s_t, s^{t-1,[m]}, z^{t-1}, u^t, n^t)\, p(s_t \mid s^{t-1,[m]}, z^{t-1}, u^t, n^t) ds_t \tag{4.29}$$

$$\stackrel{Markov}{=} \eta \int p(z_t \mid s_t, s^{t-1,[m]}, z^{t-1}, u^t, n^t)\, p(s_t \mid s^{t-1,[m]}, u^t) ds_t \tag{4.30}$$

$$= \eta \int\int p(z_t \mid \theta_{n_t}, s_t, s^{t-1,[m]}, z^{t-1}, u^t, n^t) \tag{4.31}$$
$$p(\theta_{n_t} \mid s_t, s^{t-1,[m]}, z^{t-1}, u^t, n^t) d\theta_{n_t}\, p(s_t \mid s^{t-1,[m]}, u^t) ds_t$$

$$\stackrel{Markov}{=} \eta \int\int \underbrace{p(z_t \mid \theta_{n_t}, s_t, n_t)}_{\sim\mathcal{N}(z_t; g(\theta_{n_t}, s_t), R_t)}\ \underbrace{p(\theta_{n_t} \mid s^{t-1,[m]}, z^{t-1}, u^{t-1}, n^{t-1})}_{\sim\mathcal{N}(\theta_{n_t}; \mu_{n_t,t-1}^{[m]}, \Sigma_{n_t,t-1}^{[m]})} d\theta_{n_t}$$
$$\underbrace{p(s_t \mid s^{t-1,[m]}, u^t)}_{\sim\mathcal{N}(s_t; \hat{s}_t, P_t)} ds_t \tag{4.32}$$

The three terms in this expression are all Gaussians, corresponding to the measurement model, the landmark estimate at $t-1$, and the motion model.

Using the linearization of g, this expression can be calculated in closed form. Two applications of the convolution theorem yields:

$$w_t^{[m]} \sim N(z_t; \hat{z}_t, \underbrace{G_s P_t G_s^T + G_\theta \Sigma_{n_t,t-1}^{[m]} G_\theta^T + R_t}_{L_t}) \tag{4.33}$$

Put differently, the (non-normalized) importance weight for the m-th particle is given by the following expression:

$$L_t = G_s P_t G_s^T + G_\theta \Sigma_{n_t,t-1}^{[m]} G_\theta^T + R_t \tag{4.34}$$

$$w_t^{[m]} = |2\pi L_t^{[m]}|^{-\frac{1}{2}} \exp\left\{ -\frac{1}{2}(z - \hat{z}_t^{[m]})^T L_t^{[m],-1}(z - \hat{z}_t^{[m]}) \right\} \tag{4.35}$$

4.2.3 FastSLAM 2.0 Overview

The FastSLAM 2.0 algorithm differs from the original algorithm only in the choice of proposal distributions and the calculation of the importance weights. The rest of the algorithm, including the landmark updates, data association, and resampling procedures, remain unchanged. The complete FastSLAM 2.0 algorithm is shown in Table 4.1.

4.2.4 Handling Simultaneous Observations

Simultaneous observations can be incorporated into FastSLAM 1.0 by processing each observation sequentially. Each landmark filter is updated separately and the weight of the resulting particle is the product of the weights of each observation handled individually. Incorporating simultaneous observations into FastSLAM 2.0 is slightly more complicated because the observations must also be incorporated into the proposal distribution. Instead of throwing the proposal distribution away after drawing a sample, the proposal distribution must be kept and updated for each observation. As each observation is incorporated, the proposal distribution will shrink. A sample is drawn from the incremental proposal distribution after incorporating each observation in order to update the landmark filters and compute the importance weights.

The initial proposal distribution is set to a Gaussian with mean $\hat{s}_t$ and covariance P_t. Then the proposal distribution is updated as follows:

$$\Sigma_{s_t,0} = P_t \tag{4.36}$$

$$\mu_{s_t,0} = \hat{s}_t \tag{4.37}$$

$$\Sigma_{s_t,n} = \left[G_{s,n}^T Z_{t,n}^{-1} G_{s,n} + \Sigma_{s_t,n-1}^{-1} \right]^{-1} \tag{4.38}$$

$$\mu_{s_t,n} = \mu_{s_t,n-1} + \Sigma_{s_t,n} G_{s_t,n}^T Z_{t,n}^{-1}(z_t - \hat{z}_{t,n}) \tag{4.39}$$

The creation of new landmarks in FastSLAM 2.0 deserves further consideration when multiple observations are incorporated per time step. If the first of

Table 4.1. The FastSLAM 2.0 algorithm.

Algorithm FastSLAM 2.0$(S_{t-1}, z_t, R_t, u_t, P_t)$
  $S_t = S_{aux} = \emptyset$
  for $m = 1$ **to** M // loop over all particles
    retrieve m-th particle
      $\left\langle s_{t-1}^{[m]}, N_{t-1}^{[m]}, \mu_{1,t-1}^{[m]}, \Sigma_{1,t-1}^{[m]}, \ldots, \mu_{N_{t-1}^{[m]},t-1}^{[m]}, \Sigma_{N_{t-1}^{[m]},t-1}^{[m]} \right\rangle$ from S_{t-1}
    for $n = 1$ **to** $N_{t-1}^{[m]}$ // loop over potential data associations
      $\hat{z}_{t,n} = g(\hat{s}_t, \mu_{n,t-1}^{[m]})$
      $G_{s,n} = \nabla_{s_t} g(s_t, \theta_n)|_{s_t=\hat{s}_t;\theta_n=\mu_{n,t-1}^{[m]}}$
      $Z_{t,n} = R_t + G_{\theta,n} \Sigma_{n,t-1}^{[m]} G_{\theta,n}^T$
      $\Sigma_{s_t,n} = \left[G_{s,n}^T Z_{t,n}^{-1} G_{s,n} + P_t^{-1} \right]^{-1}$
      $\mu_{s_t,n} = \hat{s}_t + \Sigma_{s_t,n}^{[m]} G_{s,n}^T Z_{t,n}^{[m]-1} (z_t - \hat{z}_{t,n})$
      $s_{t,n}^{[m]} \sim \mathcal{N}(s_t; \mu_{s_t,n}, \Sigma_{s_t,n})$
      $p_n = |2\pi Z_{t,n}|^{-\frac{1}{2}} \exp\left\{ -\frac{1}{2}(z_t - g(s_{t,n}^{[m]}, \mu_{n,t-1}^{[m]}))^T Z_{t,n}^{-1} (z_t - g(s_{t,n}^{[m]}, \mu_{n,t-1}^{[m]})) \right\}$
    end for
    $p_{N_{t-1}^{[m]}+1}^{[m]} = p_0$ // likelihood of new feature
    $\hat{n}_t = \text{argmax}_n\, p_n$ or draw $\hat{n}_t$ with prob. $\propto p_n$ // pick a data association
    if $\hat{n}_t = N_{t-1}^{[m]} + 1$ // new feature?
      $N_t^{[m]} = N_{t-1}^{[m]} + 1$
      $G_{\theta,\hat{n}_t} = \nabla_{\theta_n} g(s_t, \theta_n)|_{s_t=s_t^{[m]};\theta_n=\mu_{n,t-1}^{[m]}}$
      $\mu_{\hat{n}_t,t}^{[m]} = g^{-1}(s_t^{[m]}, \hat{z}_{\hat{n}_t,t})$ $\Sigma_{\hat{n}_t,t}^{[m]} = \left(G_{\theta,\hat{n}_t}^T R^{-1} G_{\theta,\hat{n}_t} \right)^{-1}$
      $w_t^{[m]} = p_0$
    else // known feature?
      $s_t^{[m]} = s_{t,\hat{n}_t}^{[m]}$ $N_t^{[m]} = N_{t-1}^{[m]}$
      $K_{\hat{n}_t,t} = \Sigma_{\hat{n}_t,t-1} G_{\theta,\hat{n}_t}^T Z_{\hat{n}_t,t}^{-1}$
      $\mu_{\hat{n}_t,t}^{[m]} = \mu_{\hat{n}_t,t-1}^{[m]} + K_{\hat{n}_t,t}(z_t - \hat{z}_{\hat{n}_t,t})$ $\Sigma_{\hat{n}_t,t} = (I - K_{\hat{n}_t,t} G_{\theta,\hat{n}_t}) \Sigma_{\hat{n}_t,t-1}^{[m]}$
      $L_t = G_{s,\hat{n}_t} P_t G_{s,\hat{n}_t}^T + G_{\theta,\hat{n}_t} \Sigma_{\hat{n}_t,t-1}^{[m]} G_{\theta,\hat{n}_t}^T + R_t$
      $w_t^{[m]} = |2\pi L_t|^{-\frac{1}{2}} \exp\left\{ -\frac{1}{2}(z_t - \hat{z}_{t,\hat{n}_t})^T L_t^{-1} (z_t - \hat{z}_{t,\hat{n}_t}) \right\}$
    end if
    for $n = 1$ **to** $N_t^{[m]}$ // unobserved features
      if $n \neq \hat{n}_t$
        $\Sigma_{\theta_n,t}^{[m]} = \Sigma_{\theta_n,t-1}^{[m]}$
      end if
    end for
    add $\left\langle s_t^{[m]}, N_t^{[m]}, \mu_{1,t}^{[m]}, \Sigma_{1,t}^{[m]}, \ldots, \mu_{N_t^{[m]},t}^{[m]}, \Sigma_{N_t^{[m]},t}^{[m]}, w_t^{[m]} \right\rangle$ **to** S_{aux}
  end for
  for $m = 1$ **to** M // resample M new particles
    draw particle from S_{aux} with prob. $\propto w_t^{[m]}$
    add new particle **to** S_t
  end for
  return S_t

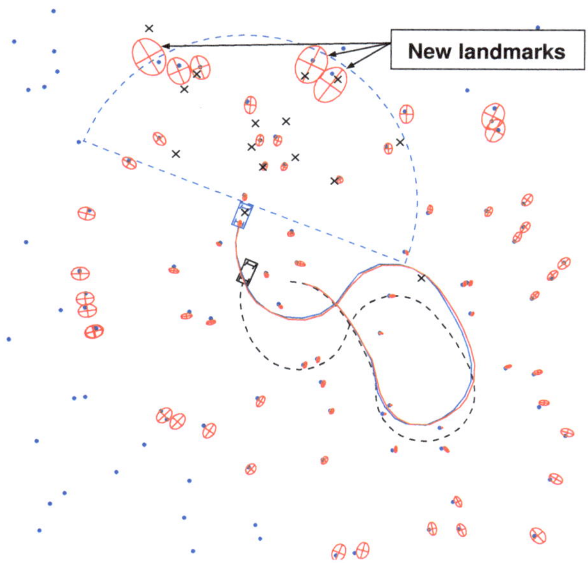

Fig. 4.3. New landmarks appear at the leading edge of the robot's field of view.

K simultaneous observations predicts the addition of a new landmark, the new robot pose $s_t^{[m]}$ will be drawn from the standard motion model $p(s_t \mid s_{t-1}, u_t)$. If this motion model is noisy, then the initial position of the new landmark may be quite inaccurate. If the observation that generates a new landmark is processed last out of the K observations, however, then the new robot pose $s_t^{[m]}$ will be drawn from the modified proposal distribution $\mathcal{N}(\theta_n, \mu_{s_t,n}, \Sigma_{s_t,n})$, which has been refined by the previous $K - 1$ observations. As a result, the initial position of the landmark will be much more accurate.

As a general rule, if multiple observations are incorporated per time step, observations that lead to new landmarks should always be processed last. One approach to this problem is to process the observations in two passes. Observations that are attributed to existing landmarks are processed in the first pass, while observations creating new landmarks are handled in the second pass. Knowledge about the robot's configuration can also be used to order the observations in such a way that new landmarks are processed last. For example, observations that generate new landmarks typically occur at the leading edge of a robot's field-of-view. (See Figure 4.3.) Incorporating observations into FastSLAM in order of ascending range typically incorporates observations of new landmarks last if forward looking sensors are used. A reasonable ordering of the observations will ensure the best possible accuracy in the location of new landmarks. This will result in better accuracy of the overall maps generated by FastSLAM 2.0

4.3 FastSLAM 2.0 Convergence

As experimental results will demonstrate, FastSLAM 2.0 decreases the number of particles necessary to achieve a given level of accuracy. This is especially true if the the robot has relatively accurate sensors and inaccurate motion, a situation common in the real world. FastSLAM 2.0 also has the unusual property that it can converge with a single particle. Since resampling is not needed, FastSLAM 2.0 with a single particle is a constant time SLAM algorithm. The time required to incorporate an observation is unaffected by the size of the map.

In standard particle filtering, resampling is the only phase of the algorithm that draws the sample set into agreement with the observations. If there is only one particle, resampling cannot change its trajectory, and the filter will diverge. In FastSLAM 2.0, the proposal distribution incorporates the observations as well. With a single particle, the motion model of FastSLAM 2.0 can minimize the error of the map by proposing a new robot pose that agrees with the latest observations.

In this section, we will demonstrate the convergence of FastSLAM 2.0 with a single particle for a special case of the SLAM problem – linear Gaussian SLAM (LG-SLAM). Convergence with a single particle trivially implies convergence with M particles. This proof has two important consequences. To our knowledge, the best previous SLAM algorithm for which convergence was shown requires quadratic update time [19]. This proof demonstrates convergence for a constant time algorithm. Second, this proof demonstrates that maintaining the entire covariance matrix is not a necessary prerequisite for convergence. FastSLAM 2.0 with a single particle maintains no correlation information between landmarks.

The convergence result presented here applies to linear SLAM problems with Gaussian noise. LG-SLAM problems are defined as possessing motion and measurement models of the following form

$$g(s_t, \theta_{n_t}) = \theta_{n_t} - s_t \tag{4.40}$$

$$h(u_t, s_{t-1}) = u_t + s_{t-1} \tag{4.41}$$

The LG-SLAM framework can be thought of as a robot operating in a Cartesian space equipped with a noise free compass, and sensors that measure distances to features along the coordinate axes. In this scenario, the mapping between observations and landmarks is assumed to be known.

While LG-SLAM is too restrictive to be of practical significance, it plays an important role in the SLAM literature. In particular, the Kalman filter has been shown to converge to the true map in the linear-Gaussian case [18, 69]. The Kalman filter converges to a state in which all map features are fully correlated. If the location of one feature is known, the rest of the features locations are recovered asymptotically by the filter.

The main convergence property for FastSLAM is the following:

Theorem: *Linear-Gaussian FastSLAM converges in expectation to the correct map with $M = 1$ particle if all features are observed infinitely often, and the location of one feature is known in advance.*

If no features are known in advance, the map will be correct in relative terms, up to a fixed offset that is applied uniformly to all features.

4.3.1 Convergence Proof

Before the proof is given, we must reformulate the FastSLAM update equations for the LG-SLAM problem using the definitions of g and h given in (4.40) and (4.41). In particular, the equations for the computation of the proposal distribution can be rewritten as follows:

$$\hat{s}_t = h(s_{t-1}^{[m]}, u_t) = s_{t-1}^{[m]} + u_t \tag{4.42}$$

$$\hat{z}_t = g(\hat{s}_t, \mu_{n_t,t-1}^{[m]}) = \mu_{n_t,t-1}^{[m]} - s_{t-1}^{[m]} - u_t \tag{4.43}$$

$$G_\theta = \nabla_{\theta_n} g(s_t, \theta_n)\big|_{s_t=\hat{s}_t;\theta_n=\mu_{n_t,t-1}^{[m]}} = I \tag{4.44}$$

$$G_s = \nabla_{s_t} g(s_t, \theta_n)\big|_{s_t=\hat{s}_t;\theta_n=\mu_{n_t,t-1}^{[m]}} = -I \tag{4.45}$$

$$Z_t = R_t + G_\theta \Sigma_{n_t,t-1}^{[m]} G_\theta^T = R_t + \Sigma_{n_t,t-1}^{[m]} \tag{4.46}$$

$$\Sigma_{s_t} = \left[G_s^T Z_t G + P_t^{-1}\right]^{-1} = \left[(R_t + \Sigma_{n_t,t-1}^{[m]})^{-1} + P_t^{-1}\right]^{-1} \tag{4.47}$$

$$\mu_{s_t} = \Sigma_{s_t} G_s^T Z_t^{-1}(z_t - \hat{z}_t) + \hat{s}_t \tag{4.48}$$

$$= -\Sigma_{s_t}(R_t + \Sigma_{n_t,t-1}^{[m]})^{-1}(z_t - \mu_{n_t,t-1}^{[m]} + s_{t-1}^{[m]} + u_t) + s_{t-1}^{[m]} + u_t \tag{4.49}$$

$$s_t^{[m]} \sim \mathcal{N}(s_t; \mu_{s_t}, \Sigma_{s_t}) \tag{4.50}$$

Similarly, the equations for updating the landmark position estimates (3.29) though (3.34) can be rewritten and consolidated:

$$\mu_{n_t,t} = \mu_{n_t,t-1} + \Sigma_{n_t,t-1}(R_t + \Sigma_{n_t,t-1}^{[m]})^{-1}(z_t - \mu_{n_t,t-1}^{[m]} + s_{t-1}^{[m]} + u_t) \tag{4.51}$$

$$\Sigma_{n_t,t} = (I - \Sigma_{n_t,t-1}^{[m]}(R_t + \Sigma_{n_t,t-1}^{[m]})^{-1})\Sigma_{n_t,t-1}^{[m]} \tag{4.52}$$

The equation for the computation of the importance weights can be ignored because this proof is for $M = 1$ particle. The resampling step of the particle filter is no longer applicable. For the proof, it will be convenient to introduce error variables $\alpha_t^{[m]}$ and $\beta_{n,t}^{[m]}$, for the robot pose and the feature locations, respectively.

$$\alpha_t^{[m]} = s_t^{[m]} - s_t \tag{4.53}$$

$$\beta_{n,t}^{[m]} = \mu_{n,t}^{[m]} - \theta_n \tag{4.54}$$

These variables measure the absolute error of the robot and the landmarks at time t. We will refer to the landmark with the known position as the *anchoring feature*. Without loss of generality, we will assume that the anchoring feature is θ_1.

The FastSLAM convergence proof is carried out through a series of lemmas. The first such lemma characterizes the effect of map errors β on the pose error α.

Lemma 1: *If the error $\beta_{n_t,t}^{[m]}$ of the observed feature z_t is smaller in magnitude than the robot pose error $\alpha_t^{[m]}$, $\alpha_t^{[m]}$ shrinks in expectation as a result of the measurement. Conversely, if $\beta_{n_t,t}^{[m]}$ is larger than $\alpha_t^{[m]}$, the latter may increase, but in expectation will not exceed $\beta_{n_t,t}^{[m]}$.*

Proof of Lemma 1: The expected error of the robot pose sample at time t is given by

$$E[\alpha_t^{[m]}] = E[s_t^{[m]} - s_t] = E[s_t^{[m]}] - E[s_t] \tag{4.55}$$

The first term can be calculated via the sampling distribution (4.50), and the second term is obtained from the linear motion model (4.41):

$$\begin{aligned} E[\alpha_t^{[m]}] =& E[-\Sigma_{s_t}(R_t + \Sigma_{n_t,t-1}^{[m]})^{-1}(z_t - \mu_{n_t,t-1}^{[m]} + s_{t-1}^{[m]} + u_t) + s_{t-1}^{[m]} + u_t] \\ & -E[u_t + s_{t-1}] \qquad (4.56) \\ =& -\Sigma_{s_t}(R_t + \Sigma_{n_t,t-1}^{[m]})^{-1}(E[z_t] - \mu_{n_t,t-1}^{[m]} + s_{t-1}^{[m]} + u_t) + \underbrace{s_{t-1}^{[m]} - s_{t-1}}_{\alpha_{t-1}^{[m]}} \qquad (4.57) \end{aligned}$$

The last transformation exploits the linearity of the expectation. We note that in LG-SLAM the expectation $E[z_t] = \theta_{n_t} - E[s_t] = \theta_{n_t} - u_t - s_{t-1}$. With that, the expression in parentheses becomes:

$$\begin{aligned} E[z_t] - \mu_{n_t,t-1}^{[m]} + s_{t-1}^{[m]} + u_t &= \theta_{n_t} - u_t - s_{t-1} - \mu_{n_t,t-1}^{[m]} + s_{t-1}^{[m]} + u_t \qquad (4.58) \\ &= s_{t-1}^{[m]} - s_{t-1} + \theta_{n_t} - \mu_{n_t,t-1}^{[m]} \qquad (4.59) \\ &= \alpha_{t-1}^{[m]} - \beta_{n_t,t-1}^{[m]} \qquad (4.60) \end{aligned}$$

Plugging this back into (4.57) and substituting $\Sigma_{s_t}^{[m]}$ from (4.47) gives us:

$$\begin{aligned} E[\alpha_t^{[m]}] =& \alpha_{t-1}^{[m]} + \Sigma_{s_t}(R_t + \Sigma_{n_t,t-1}^{[m]})^{-1}(\beta_{n_t,t-1}^{[m]} - \alpha_{t-1}^{[m]}) \qquad (4.61) \\ =& \alpha_{t-1}^{[m]} + \left[(R_t + \Sigma_{n_t,t-1}^{[m]})^{-1} + P_t^{-1}\right]^{-1}(R_t + \Sigma_{n_t,t-1}^{[m]})^{-1}(\beta_{n_t,t-1}^{[m]} - \alpha_{t-1}^{[m]}) \qquad (4.62) \\ =& \alpha_{t-1}^{[m]} + \left[I + (R_t + \Sigma_{n_t,t-1}^{[m]})P_t^{-1}\right]^{-1}(\beta_{n_t,t-1}^{[m]} - \alpha_{t-1}^{[m]}) \qquad (4.63) \end{aligned}$$

R_t, $\Sigma^{[m]}_{n_t,t-1}$, and P_t^{-1} are all positive semidefinite, so the inverse of $I + (R_t + \Sigma^{[m]}_{n_t,t-1})P_t^{-1}$ will also be positive semidefinite, with all eigenvalues being less than one. This observation effectively proves lemma 1. In particular, the expected pose error $\alpha_t^{[m]}$ shrinks if $\beta_{n_t,t-1}$ is smaller in magnitude than $\alpha_{t-1}^{[m]}$. Conversely, if $\alpha_{t-1}^{[m]}$ is smaller in magnitude than $\beta^{[m]}_{n_t,t-1}$, Equation (4.63) suggests that $\alpha_t^{[m]}$ will increase in expectation, but only by a value that is proportional to the difference. This ensures that $\alpha_t^{[m]}$ will not exceed $\beta^{[m]}_{n_t,t}$ in expectation. *qed.*

Lemma 2: *If the robot observes the anchoring feature, the pose error of the robot will shrink in expectation.*

Proof of Lemma 2: For the anchoring feature θ_1, we can exploit the fact that $\Sigma^{[m]}_{1,t} = \beta^{[m]}_{1,t} = 0$. The lemma now follows directly from equation (4.63),

$$E[\alpha_t^{[m]}] = \alpha_{t-1}^{[m]} + \left[I + (R_t + 0)P_t^{-1}\right]^{-1}(0 - \alpha_{t-1}^{[m]}) \tag{4.64}$$

$$= \alpha_{t-1}^{[m]} - \left[I + R_t P_t^{-1}\right]^{-1}\alpha_{t-1}^{[m]} \tag{4.65}$$

Thus, whenever the robot sees its anchoring feature, its pose error $\alpha_t^{[m]}$ is guaranteed to shrink. The only exception arises when the error is already zero, in which case it remains zero in expectation. *qed.*

Finally, a lemma similar to Lemma 1 states the effect of pose errors α on map errors β.

Lemma 3: *If the pose error $\alpha_{t-1}^{[m]}$ is smaller in magnitude than the error $\beta^{[m]}_{n_t,t-1}$ of the observed feature, observing z_t shrinks $\beta^{[m]}_{n_t,t}$ in expectation. Conversely, if $\alpha_{t-1}^{[m]}$ is larger than feature error $\beta^{[m]}_{n_t,t-1}$, the latter may increase, but will never exceed $\alpha_{t-1}^{[m]}$ in expectation.*

Proof of Lemma 3: This proof is analogous to that of Lemma 1. From (4.51) it follows that the expected feature error after the landmark update is:

$$E[\beta^{[m]}_{n_t,t}] = E[\mu^{[m]}_{n_t,t} - \theta_{n_t}] = E[\mu^{[m]}_{n_t,t}] - \theta_{n_t} \tag{4.66}$$

$$= E[\mu^{[m]}_{n_t,t-1} + \Sigma^{[m]}_{n_t,t-1}(R_t + \Sigma^{[m]}_{n_t,t-1})^{-1}(z_t - \mu^{[m]}_{n_t,t-1} + s_{t-1}^{[m]} + u_t)] - \theta_{n_t} \tag{4.67}$$

$$=\mu^{[m]}_{n_t,t-1} + \Sigma^{[m]}_{n_t,t-1}(R_t + \Sigma^{[m]}_{n_t,t-1})^{-1}(E[z_t] - \mu^{[m]}_{n_t,t-1} + s^{[m]}_{t-1} + u_t) - \theta_{n_t} \tag{4.68}$$

Equation (4.60) enables us to rewrite this equation as follows:

$$E[\beta^{[m]}_{n_t,t}] = \mu^{[m]}_{n_t,t-1} + \Sigma^{[m]}_{n_t,t-1}(R_t + \Sigma^{[m]}_{n_t,t-1})^{-1}(\alpha^{[m]}_{t-1} - \beta^{[m]}_{n_t,t-1}) - \theta_{n_t} \tag{4.69}$$

$$= \beta^{[m]}_{n_t,t-1} + \Sigma^{[m]}_{n_t,t-1}(R_t + \Sigma^{[m]}_{n_t,t-1})^{-1}(\alpha^{[m]}_{t-1} - \beta^{[m]}_{n_t,t-1}) \tag{4.70}$$

$$= \beta^{[m]}_{n_t,t-1} + (I + R_t\Sigma^{[m]-1}_{n_t,t-1})^{-1}(\alpha^{[m]}_{t-1} - \beta^{[m]}_{n_t,t-1}) \tag{4.71}$$

Again, since $\Sigma^{[m]}_{n_t,t-1}$ and R_t are both positive semidefinite, the eigenvalues of $(I + R_t\Sigma^{[m]-1}_{n_t,t-1})^{-1}$ are all positive and less than one, which proves the lemma. *qed.*

Proof of the Convergence Theorem: Let $\hat{\beta}^{[m]}_t$ denote the largest feature error at time t.

$$\hat{\beta}^{[m]}_t = \underset{\beta^{[m]}_{n,t}}{\operatorname{argmax}} |\beta^{[m]}_{n,t}| \tag{4.72}$$

Lemma 3 says that this error may increase in expectation, but only if the absolute robot pose error $\alpha^{[i]}_{t-1}$ exceeds this error in magnitude. However, in expectation this will only be the case for a limited number of iterations. In particular, Lemma 1 guarantees that $\alpha^{[m]}_{t-1}$ may only shrink in expectation if this is the case. Furthermore, Lemma 2 states that every time the anchoring feature is observed, this error will shrink by a finite amount, regardless of the magnitude of $\hat{\beta}^{[m]}_t$. Hence, $\alpha^{[m]}_{t-1}$ will ultimately become smaller in magnitude (again in expectation) than the largest feature error. Once this has happened, Lemma 3 states that the latter will shrink in expectation every time the feature is observed whose error is largest. It is now easy to see that both $\hat{\beta}^{[m]}_t$ and $\alpha^{[m]}_t$ converge to zero; observing the anchoring feature induces a finite reduction as stated in (4.65). To increase $\alpha^{[m]}_{t-1}$ to its old value in expectation, the total feature error must shrink in expectation according to (4.71). This leads to an eternal shrinkage of the total feature error down to zero. Since this error is an upper bound for the expected pose error (see Lemma 1), we also have convergence in expectation for the robot pose error. *qed.*

Theorem 1 trivially implies the following corollary, which characterizes the convergence of FastSLAM in the linear Gaussian setting with more than one particle.

Corollary 1: *FastSLAM converges in expectation for LG-SLAM if all features are observed infinitely often and the location of one feature is known in advance.*

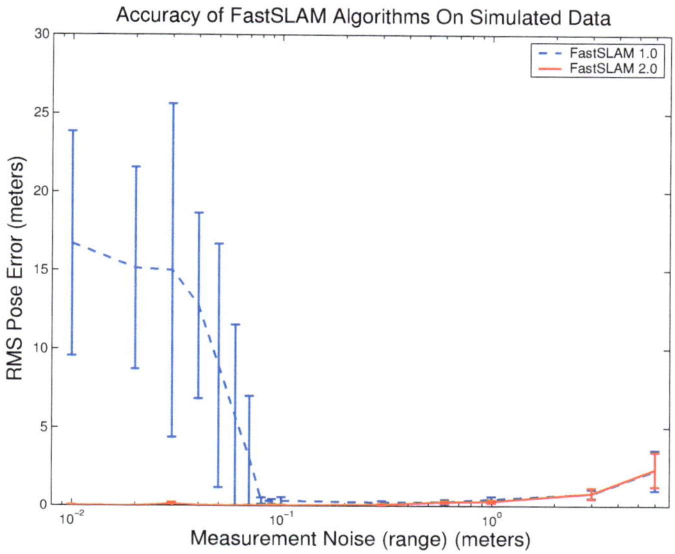

Fig. 4.4. FastSLAM 1.0 and 2.0 with varying levels of measurement noise.

4.4 Experimental Results

4.4.1 FastSLAM 1.0 Versus FastSLAM 2.0

In general, the performance of FastSLAM 2.0 will be similar to the performance of FastSLAM 1.0. However, in situations where the measurement error is significantly small compared to the motion error, FastSLAM 2.0 will outperform the original algorithm. In the following experiment, the performance of the two algorithms is compared on simulated data while varying the level of measurement noise. All experiments were run with 100 particles and known data association. The range and bearing error parameters were scaled equally.

The results of this experiment are shown in Figure 4.4. As the measurement error gets very large, the errors of both FastSLAM 1.0 and 2.0 begin to increase slowly, as expected. In this range, the performance of the two algorithms is roughly equal. For very low values of measurement error, FastSLAM 1.0 clearly begins to diverge, while the error of FastSLAM 2.0 continues to shrink. By adding more particles, the threshold below which FastSLAM 1.0 diverges can be decreased. However, FastSLAM 2.0 can produce accurate maps in these situations without increasing the number of particles.

The performance of the two algorithms can also be compared by keeping the measurement model constant and varying the number of particles. FastSLAM 2.0 will require fewer particles than FastSLAM 1.0 in order to achieve a given level of accuracy, especially when measurement error is low. In the limit, FastSLAM 2.0 can produce reasonable maps with just a single particle,

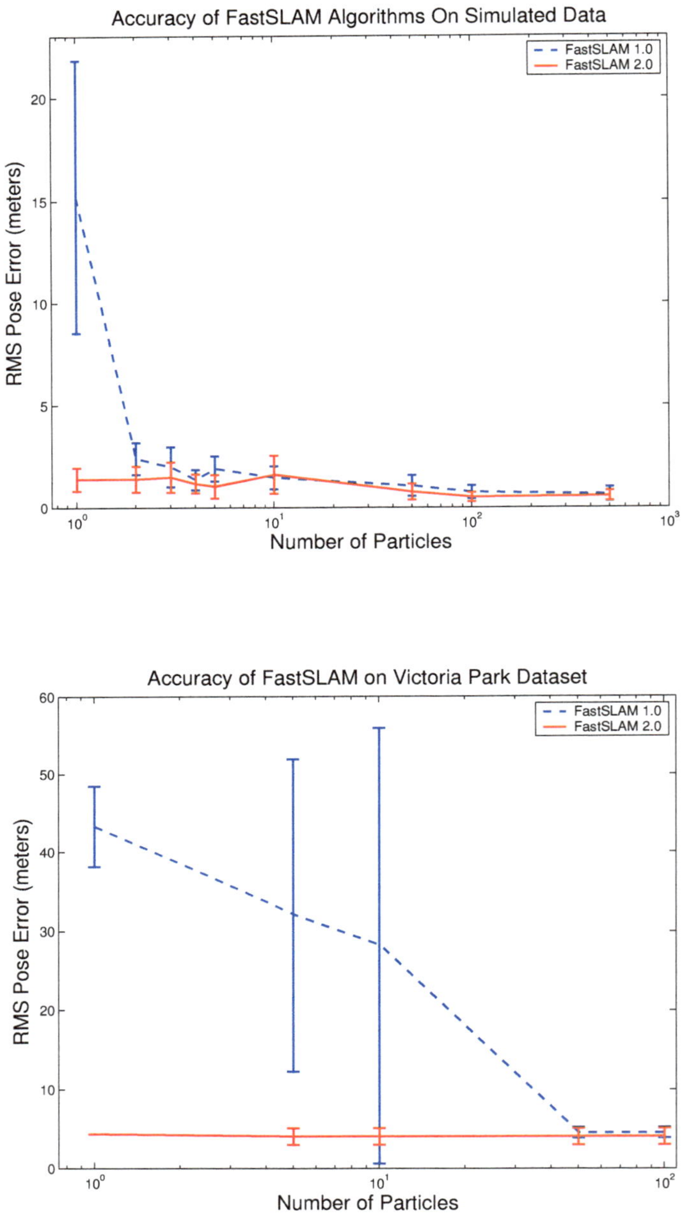

Fig. 4.5. Performance of FastSLAM algorithms with different numbers of particles.

Fig. 4.6. Map of Victoria Park make with FastSLAM 2.0 with $M = 1$ particle.

while FastSLAM 1.0 will obviously diverge. Figure 4.5 shows the results of an experiment comparing the performance of FastSLAM 1.0 and 2.0 given different numbers of particles. The two algorithms were run repeatedly on simulated data and the Victoria Park data set. On the simulated data, the accuracy of the two algorithms is similar with more than five particles. Below five particles, FastSLAM 1.0 begins to diverge and the performance of FastSLAM 1.0 stays approximately constant.

On the Victoria Park dataset the difference between the two algorithms is even more pronounced. Below 50 particles, FastSLAM 1.0 starts to diverge. Again, this is because the vehicle's controls are noisy relative to the sensor observations.

4.4.2 One Particle FastSLAM 2.0

The data associations in the Victoria Park data set are relatively unambiguous, so the one particle version of FastSLAM 2.0 can be used. With only a single particle, data association in FastSLAM 2.0 is equivalent to the maximum likelihood data association algorithm of the EKF. Figure 4.6 shows the output of FastSLAM with a single particle. The algorithm is able to produce results on par with those of the EKF and FastSLAM 1.0 without storing any correlations between landmarks.

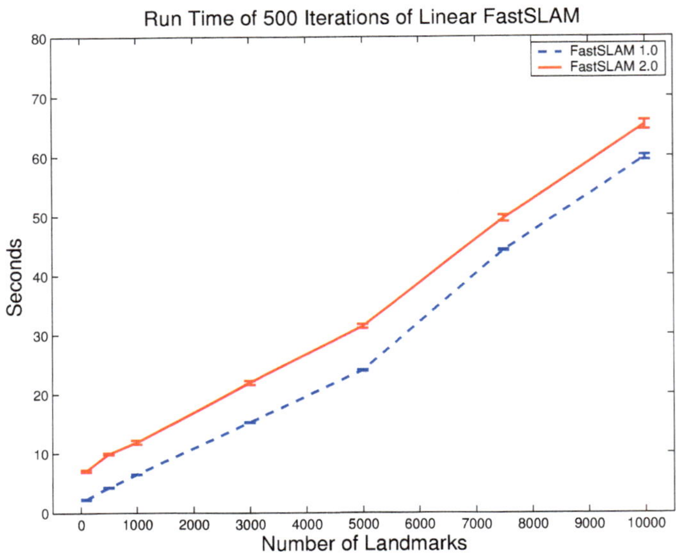

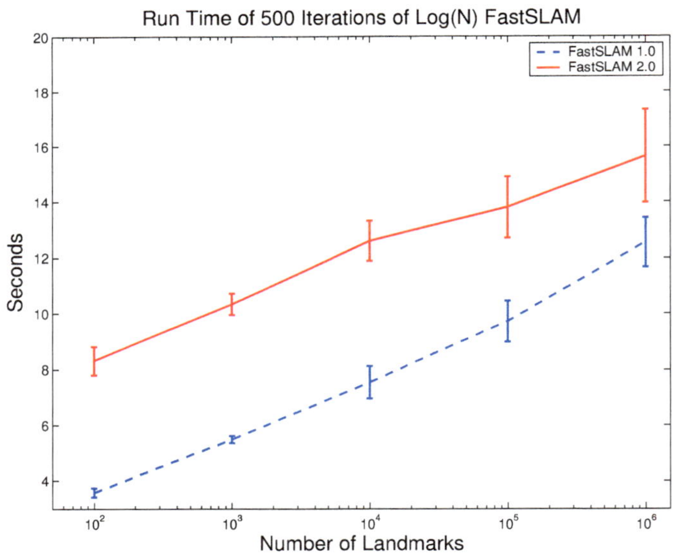

Fig. 4.7. Comparison of FastSLAM 1.0 and FastSLAM 2.0 timing.

4.4.3 Scaling Performance

The experiment in Section 4.4.1 demonstrates that FastSLAM 2.0 requires fewer particles than FastSLAM 1.0 in order to achieve a given level of estimation accuracy. Fewer particles, in turn, results in faster sensor updates. However, the construction of the improved proposal distribution requires extra time over the FastSLAM 1.0 proposal. As the number of landmarks in the map increases, the sensor updates take a smaller fraction of the overall run time relative to the importance resampling. In larger maps, the large savings gained as a result of needing fewer particles overwhelms the additional complexity of drawing from the proposal distribution. The actual savings will depend on the parameters of the motion and measurement models.

Figure 4.7 shows the run time for the linear and $\log(N)$ versions of FastSLAM 1.0 and 2.0 all with 100 particles. In very small maps (i.e. 100 landmarks), FastSLAM 2.0 requires approximately 3 times longer to perform a sensor update. However, in larger maps the sensor updates only require 10-20% more time. The constant difference between FastSLAM 1.0 and 2.0 with an equal number of particles depends primarily on the average number of observations incorporated per time step.

4.4.4 Loop Closing

In FastSLAM, the ability to close loops effectively depends on the number of particles M. The minimum number of particles is difficult to quantify, because it depends on a number of factors, including the parameters of the motion and measurement models and the density of landmarks in the environment. FastSLAM 2.0's improved proposal distribution insure that fewer particles are eliminated in resampling compared to FastSLAM 1.0. Better diversity in the sample set results in better loop closing performance, because new observations can affect the pose of the robot further back in the past.

Examples of loop closing with FastSLAM 1.0 and FastSLAM 2.0 are shown in Figure 4.8(a) and (b), respectively. The histories of all M particles are drawn for both algorithms. In Figure 4.8(a), the FastSLAM 1.0 particles share a common history part of the way around the loop. New observations can not affect the positions of landmarks observed before this threshold. In this case of FastSLAM 2.0, the algorithm is able to maintain diversity that extends back to the beginning of the loop. This is crucial for reliable loop closing and fast convergence.

Figure 4.9 shows the result of an experiment comparing the loop closing performance of FastSLAM 1.0 and 2.0. As the size of the loop increases, the error of both algorithms increases. However, FastSLAM 2.0 consistently outperforms FastSLAM 1.0. Alternately, this result can rephrased in terms of particles. FastSLAM 2.0 requires fewer particles to close a given loop than FastSLAM 1.0.

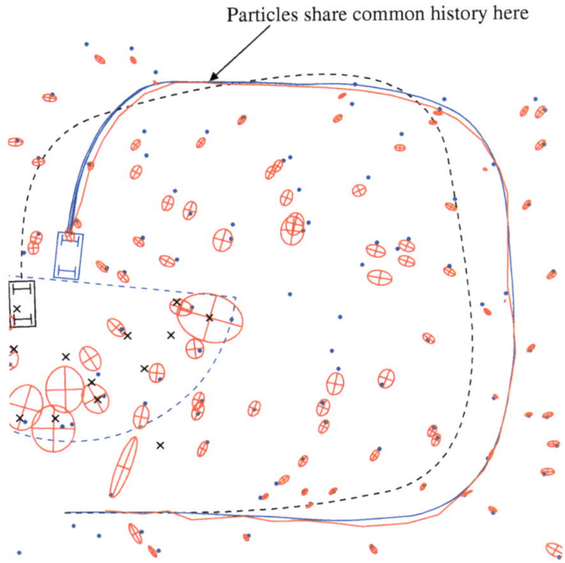

(a) FastSLAM 1.0 – Samples share a common path inside the loop. This often leads to divergence.

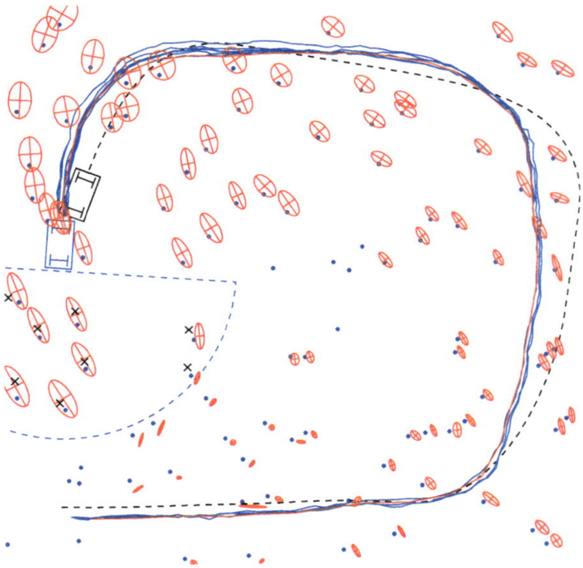

(b) FastSLAM 2.0 – The algorithm maintains path diversity all the way around the loop.

Fig. 4.8. FastSLAM 2.0 can close larger loops than FastSLAM 1.0 given a constant number of particles.

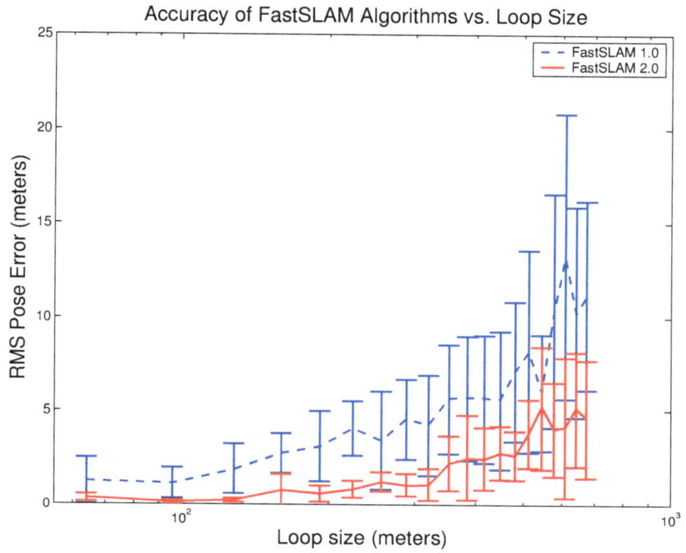

(a) Accuracy as a function of loop size

Fig. 4.9. FastSLAM 2.0 can close larger loops than FastSLAM 1.0 given a fixed number of particles.

4.4.5 Convergence Speed

By pruning away improbable trajectories of the robot, resampling eventually causes all of the FastSLAM particles to share a common history at some point in the past. New observations cannot affect the positions of landmarks observed prior to this point. This common history point can be pushed back in time by increasing the number of particles M. This process of throwing away correlation data over time enables FastSLAM's efficient sensor updates. This efficiency comes at the cost of slower convergence speed. Throwing away correlation information means that more observations will be required to achieve a given level of accuracy.

The trade-off of number of particles versus convergence speed occurs in both FastSLAM 1.0 and FastSLAM 2.0. However, FastSLAM 2.0 can operate without maintaining any cross-correlations between landmarks, so the relationship effect of throwing away correlation data on convergence speed is easier to study. In particular, this effect is most prominent when closing a large loop. Revisiting a known landmark should refine the positions of all landmarks around the loop. If correlation information is thrown away, convergence to the true map will be slower, and more trips around the loop will be necessary to achieve the same level of accuracy.

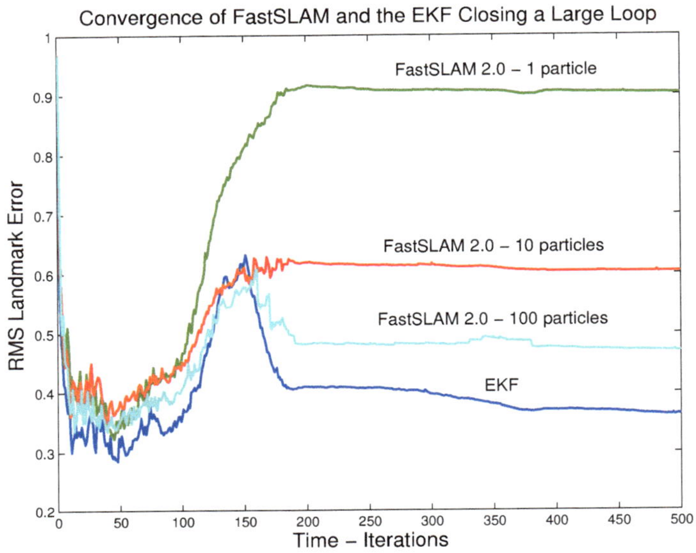

Fig. 4.10. Comparison of the convergence speed of FastSLAM 2.0 and the EKF.

Figure 4.10 shows the results of an experiment comparing the convergence speed of FastSLAM 2.0 and the EKF. FastSLAM 2.0 (with 1, 10, and 100 particles) and the EKF were each run 10 times around a large simulated loop of landmarks, similar to the ones shown in Figure 4.9(a) and (b). Different random seeds were used for each run, causing different controls and observations to be generated for each loop. The RMS position error in the map at every time step was averaged over the 10 runs for each algorithm.

As the robot goes around the loop, error should gradually build up in the map. When the robot closes the loop at iteration 150, revisiting old landmarks should affect the positions of landmarks all around the loop, causing the overall error in the map to decrease. This clearly happens in the EKF. FastSLAM 2.0 with a single particle has no way to affect the positions of past landmarks so there is no drop in the landmark error. As more particles are added to FastSLAM 2.0, the filter is able to apply observations to landmark positions further back in time, gradually approaching the convergence speed of the EKF. Clearly, the number of particles necessary to achieve convergence time close to the EKF will increase with the size of the loop. It is unknown at this time whether the number of particles necessary to achieve a given accuracy is polynomial or exponential in the size of the loop.

4.5 Grid-Based FastSLAM

FastSLAM has recently been extended beyond the Gaussian landmark paradigm, most notably by Hähel at al. [40] and Eliazar and Parr [26]. The Grid-based version of FastSLAM replaces the Gaussian landmark list by an occupancy grid map, as developed by Elfes and Moravec [25, 63]. The advantage of such an extension is that it does not require any predefined definition of landmarks. Instead, in can model arbitrary types of environments, using a dense, volumetric map. In fact, Murphy's original publication used grids for representing maps [66].

The extension of FastSLAM to occupancy grid maps is mathematically straightforward, and can be found in detail in the work referenced above and in [87]. In all cases, the independence assumption that underlies the FastSLAM paradigm is valid. Each particle retains a full occupancy grid map, instead of a landmark list. The update of the occupancy grid is identical to the original literature on this topic. As shown by Eliazar and Parr [26], it can be implemented efficiently using the analog of the tree-idea discussed in Chapter 3.7.

These following results are courtesy of Dirk Hähnel. His work[40] uses scan matching to define the proposal distribution, which makes it an instance of FastSLAM 2.0. Figure 4.11 shows a typical situation of the application of the grid-based FastSLAM algorithm. Shown there are three particles together with their associated maps. Each particle represents a potential trajectory of the robot, which explains why each occupancy grid map looks different. The center map is the best in terms of global consistency.

A typical map acquired with the FastSLAM algorithm is depicted in Figure 4.13. The size of this environment is 28m $\times$ 28m. The length of the robot's trajectory is 491m and the average speed was 0.19m/s. The resolution of the map is 10cm. To learn this map, as few as 500 particles were used. During the overall process the robot encountered two loops. A map calculated from pure odometry data is shown in Figure 4.12, illustrating the amount of error in the robot's odometry.

The importance of using multiple particles becomes evident in Figure 4.14, which visualizes the trajectories of the samples shortly before and after closing a loop. As the left image illustrates, the robot is quite uncertain about its position relative to the starting position, hence the wide spread of particles at the time of loop closure. However, a few resampling steps after the robot re-enters known terrain suffice to reduce the uncertainty drastically (right image).

These results illustrate the flexibilityof the FastSLAM paradigm. Some of the largest indoor maps ever have been generated with FastSLAM.

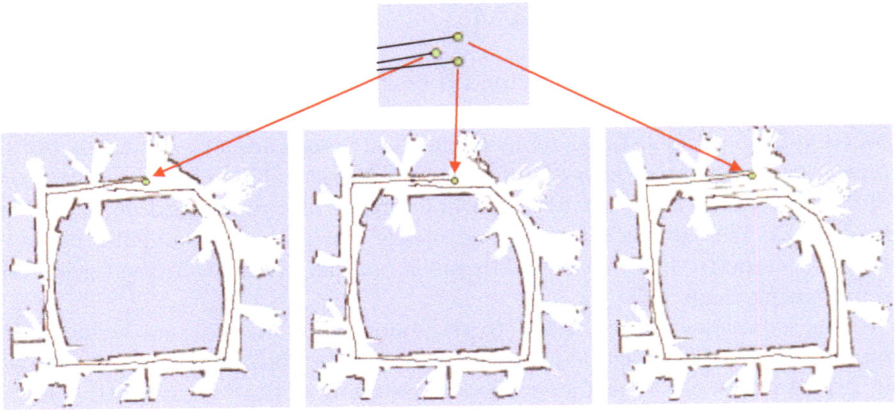

Fig. 4.11. Application of the grid-based variant of the FastSLAM algorithm. Each particle carries its own map and the importance weights of the particles are computed based on the likelihood of the measurements given the particle's own map.

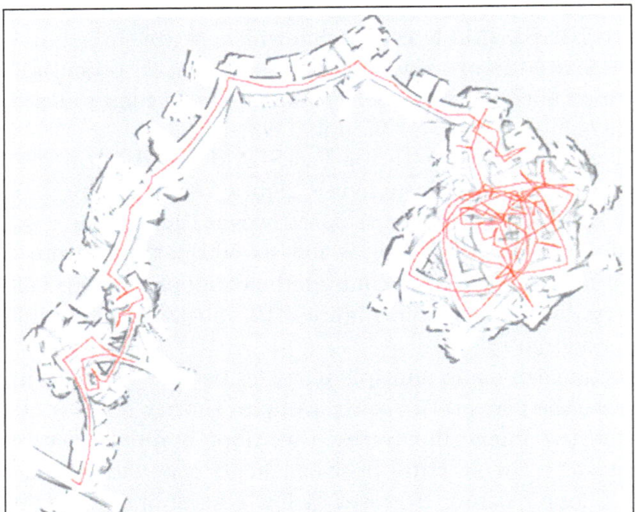

Fig. 4.12. Occupancy grid map generated from laser range data and based on pure odometry. All images courtesy of Dirk Hähnel.

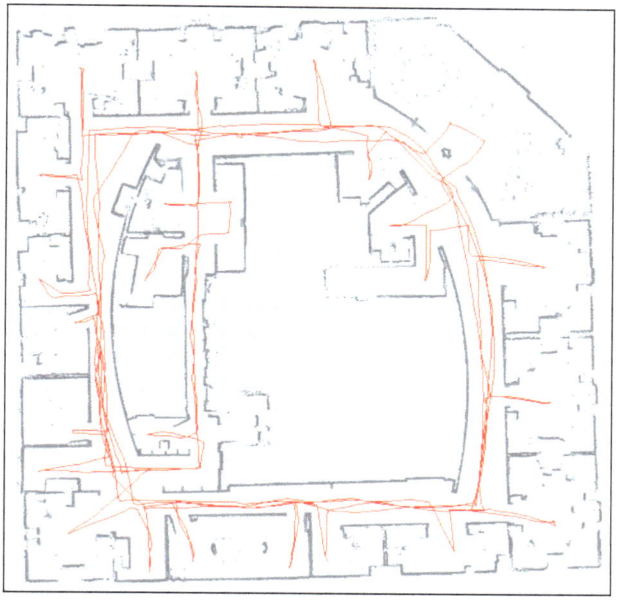

Fig. 4.13. Occupancy grid generated with the grid-based FastSLAM algorithm, using 500 particles. Also depicted in this image is the path represented by the particle with the maximum accumulated importance weight.

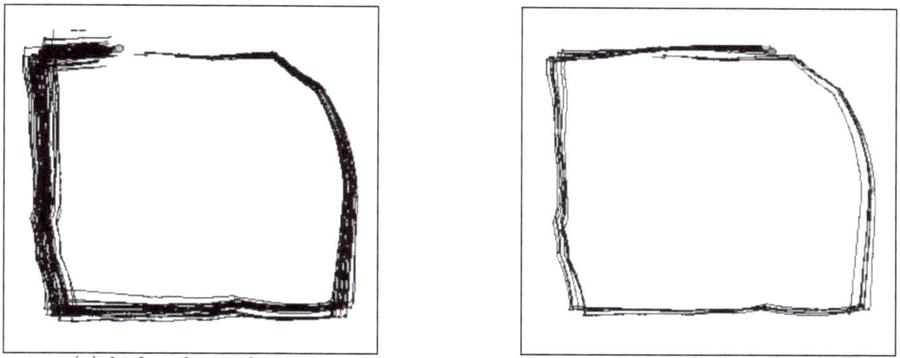

(a) before loop closure (b) after loop closure

Fig. 4.14. Trajectories of all samples shortly before (left) and after (right) closing the outer loop of the environment depicted in Figure 4.13. Images courtesy of Dirk Hähnel.

4.6 Summary

This chapter described a modified version of the FastSLAM algorithm that addresses the issue of sample impoverishment. FastSLAM 2.0 incorporates observations into the proposal distribution of the particle filter in order to better match the posterior distribution. As a consequence, fewer trajectories are thrown out during resampling, and FastSLAM 2.0 is able to maintain correlation information further back in time. This results in better performance when the robot's sensor is very accurate, faster convergence, and allows the algorithm to close larger loops. Finally, this chapter presented a convergence proof for FastSLAM 2.0 in linear-Gaussian worlds. This proof demonstrates that maintaining the full covariance is not strictly necessary for convergence. Experimental results do show, however, that eliminating correlation information can result in substantially slower convergence in practice.

5

Dynamic Environments

Thus far, this book has addressed the problem of a mobile robot navigating within a unknown, *static* map. As a result, the uncertainty in the landmark positions only decreases over time. The static world assumption is reasonable in environments like planetary and underground exploration, but it is unrealistic in typical human environment s, such as office buildings and city streets [93]. This chapter will describe a generalization of the SLAM problem to worlds with both static and dynamic objects.

Unlike static objects, the uncertainty in the position of a dynamic object increases over time if it is not observed. If all landmarks are allowed to move, then SLAM becomes an ill-posed problem. There is no fixed reference against which to bound the error of the map or the pose of the robot. Instead, by classifying objects in the world as either static or dynamic, the static objects can be used to make a map while the dynamic objects are tracked separately. This approach has been used successfully by several authors in the scan-matching community [39, 93].

Including dynamic objects requires no fundamental changes to the Fast-SLAM algorithm. The basic factorization of the SLAM problem presented in Chapter 3 applies regardless of the motion of the landmarks. However, including dynamic objects can make the SLAM problem substantially more difficult to solve in practice. In addition to solving the data association problem, the SLAM algorithm must estimate whether each landmark is static or dynamic.

After deriving the general SLAM problem with dynamic objects, we will make the strong assumption that an accurate map of the static objects is available to the robot. Mobile robots performing persistent tasks in an environment are often equipped with maps, either generated by SLAM algorithms or taken or from a priori sources such as blueprints. Instead of tracking a robot and a static map, the problem now becomes tracking a robot and dynamic features relative to a known, static map. In the experiments described in this chapter, the dynamic objects will be people. We will show how the Simultaneous Localization and People-tracking (SLAP) problem exhibits the exact same structure as the original SLAM problem.

By formulating the SLAP problem as a special case of the dynamic SLAM problem, it is evident that the factorization at the heart of FastSLAM also applies to SLAP. Consequently, the same basic algorithm can be applied to estimate the pose of the robot and the locations of the people. The primary difference between SLAM and SLAP is that the "landmarks" move over time. Thus, our algorithm must include a motion model for the people as well as the robot. Since the "controls" governing the motion of people are unknown, this motion model must be general enough to predict all possible motions of a person.

Even though SLAP is a special case of the dynamic SLAM problem, it is still an interesting and difficult problem. Uncertainty in the pose of the robot correlates error in the estimated positions of the people. Factoring the SLAP problem allows the global positions of the people to be tracked efficiently, even if the position of the robot is uncertain. Experiments in this chapter will demonstrate that a variation of the basic FastSLAM algorithm can track people reliably in situations with global uncertainty over the pose of the robot.

5.1 SLAM with Dynamic Landmarks

Chapter 2 introduced the formulation of the SLAM problem as a posterior estimation problem. The goal was to estimate a posterior over all possible maps and robot poses, given the observations, controls, and data associations.

$$p(s_t, \Theta \mid z^t, u^t, n^t) \tag{5.1}$$

In section 2.3.1, we showed how this posterior can be computed recursively as a function of the identical posterior at time $t-1$. This recursive update equation, known as the Bayes Filter, is duplicated here, for convenience.

$$\begin{aligned} &p(s_t, \Theta \mid z^t, u^t, n^t) = \\ &\eta \underbrace{p(z_t \mid s_t, \Theta, n_t)}_{\text{measurement model}} \int \underbrace{p(s_t \mid s_{t-1}, u_t)}_{\text{motion model}} \underbrace{p(s_{t-1}, \Theta \mid z^{t-1}, u^{t-1}, n^{t-1})}_{\text{posterior at time t}-1} ds_{t-1} \end{aligned} \tag{5.2}$$

Including dynamic objects in the SLAM posterior will result in a very similar recursive update equation. As in the previous chapters, the set of static features will be written as Θ. The number of static features will be written N_θ. The set of dynamic features will be written as Φ_t. The number of dynamic features will be written as N_ϕ.

$$\Theta = \{\theta_1, \theta_2, \ldots, \theta_{N_\theta}\} \tag{5.3}$$

$$\Phi_t = \{\phi_{1,t}, \phi_{2,t}, \ldots, \phi_{N_\phi,t}\} \tag{5.4}$$

In addition to the controls u_t that describe the motion of the robot, we will refer to the controls that describe the motion of the dynamic features as

$u_{\phi,t}$. In general, the robot will not have access to this information; instead we will assume generic controls $u_{\phi,t}$ that predict all possible motions of the features. The data associations n_t will now index both static and dynamic features. Each data association will describe which static or dynamic object corresponds to the observation z_t.

The posterior over robot poses and all feature locations can be written:

$$p(s_t, \Theta, \Phi_t \mid z^t, u^t, u_\phi^t, n^t) \tag{5.5}$$

The recursive update equation for this posterior can be derived in the same way as the Bayes Filter for static landmarks. This update equation can be written in two parts:

$$p(s_{t-1}, \Theta, \Phi_t \mid z^{t-1}, u^{t-1}, u_\phi^t, n^{t-1}) = \tag{5.6}$$
$$\int \underbrace{p(\Phi_t \mid \Phi_{t-1}, u_{\phi,t})}_{\text{dynamic motion model}} \, p(s_{t-1}, \Theta, \Phi_{t-1} \mid z^{t-1}, u^{t-1}, u_\phi^{t-1}, n^{t-1}) d\Phi_{t-1}$$

$$p(s_t, \Theta, \Phi_t \mid z^t, u^t, u_\phi^t, n^t) = \tag{5.7}$$
$$\underbrace{p(z_t \mid s_t, \Theta, \Phi_t, n_t)}_{\text{measurement model}} \int \underbrace{p(s_t \mid s_{t-1}, u_t)}_{\text{robot motion model}} \, p(s_{t-1}, \Theta, \Phi_t \mid z^{t-1}, u^{t-1}, u_\phi^t, n^{t-1}) ds_{t-1}$$

If there are only static objects in the world, then $p(s_{t-1}, \Theta, \Phi_t \mid z^{t-1}, u^{t-1}, u_\phi^t, n^{t-1})$ is equal to $p(s_{t-1}, \Theta \mid z^{t-1}, u^{t-1}, n^{t-1})$, the standard SLAM posterior at time $t-1$, and the measurement model becomes $p(z_t \mid s_t, \Theta, n_t)$. In this case, (5.7) reverts back to the Bayes Filter for static objects (5.2).

Similar in structure to (5.2), the posterior at time t is a function of the posterior at time $t-1$, a measurement model $p(z_t \mid s_t, \Theta, \Phi_t, n_t)$, and a robot motion model $p(s_t \mid s_{t-1}, u_t)$. The update equation also includes a second motion model $p(\Phi_t \mid \Phi_{t-1}, u_{\phi,t})$ that describes the motion of dynamic features.

5.1.1 Derivation of the Bayes Filter with Dynamic Objects

The derivation of the Bayes Filter for SLAM with dynamic features parallels the derivation for the standard Bayes Filter shown in Chapter 2. It can be safely skipped by the reader without a loss of understanding.

The first step of the derivation is to expand the posterior using Bayes Rule.

$$p(s_t, \Theta, \Phi_t \mid z^t, u^t, u_\phi^t, n^t) =$$
$$\eta\, p(z_t \mid s_t, \Theta, \Phi_t, z^{t-1}, u^t, u_\phi^t, n^t) p(s_t, \Theta, \Phi_t \mid z^{t-1}, u^t, u_\phi^t, n^t) \tag{5.8}$$

The observation z_t only depends on s_t, Θ, Φ_t, and n_t. Simplifying the first term of (5.8) reveals the new measurement model.

$$= \eta \underbrace{p(z_t \mid s_t, \Theta, \Phi_t, n_t)}_{\text{measurement model}} p(s_t, \Theta, \Phi_t \mid z^{t-1}, u^t, u_\phi^t, n^t) \tag{5.9}$$

Next, we condition the term the second term of (5.9) on s_{t-1} using the Theorem of Total Probability.

$$= \eta\, p(z_t \mid s_t \Theta, \Phi_t, n_t) \int p(s_t, \Theta, \Phi_t \mid s_{t-1}, z^{t-1}, u^t, u_\phi^t, n^t) \\ p(s_{t-1} \mid z^{t-1}, u^t, u_\phi^t, n^t) ds_{t-1} \tag{5.10}$$

The first term inside the integral can be expanded using the definition of conditional probability.

$$p(s_t, \Theta, \Phi_t \mid s_{t-1}, z^{t-1}, u^t, u_\phi^t, n^t) = \\ p(s_t \mid \Theta, \Phi_t, s_{t-1}, z^{t-1}, u^t, u_\phi^t, n^t) p(\Theta, \Phi_t \mid s_{t-1}, z^{t-1}, u^t, u_\phi^t, n^t) \tag{5.11}$$

The first term of (5.11) can be simplified by noting that s_t only depends on s_{t-1} and the robot control u_t.

$$p(s_t, \Theta, \Phi_t \mid s_{t-1}, z^{t-1}, u^t, u_\phi^t, n^t) \\ = \underbrace{p(s_t \mid s_{t-1}, u_t)}_{\text{robot motion model}} p(\Theta, \Phi_t \mid s_{t-1}, z^{t-1}, u^t, u_\phi^t, n^t) \tag{5.12}$$

By substituting (5.12) back into (5.10), and combining the rightmost terms of both equations according to the definition of conditional probability, the equation for the posterior becomes:

$$p(s_t, \Theta, \Phi_t \mid z^t, u^t, u_\phi^t, n^t) = \\ \eta\, p(z_t \mid s_t \Theta, \Phi_t, n_t) \int p(s_t \mid s_{t-1}, u_t) p(s_{t-1}, \Theta, \Phi_t \mid z^{t-1}, u^t, u_\phi^t, n^t) ds_{t-1} \tag{5.13}$$

The final term in the integral can be simplified by noting that u_t and n_t both do not affect s_{t-1} or the landmark positions.

$$p(s_t, \Theta, \Phi_t \mid z^t, u^t, u_\phi^t, n^t) = \\ \eta\, p(z_t \mid s_t \Theta, \Phi_t, n_t) \int p(s_t \mid s_{t-1}, u_t) p(s_{t-1}, \Theta, \Phi_t \mid z^{t-1}, u^{t-1}, u_\phi^t, n^{t-1}) ds_{t-1} \tag{5.14}$$

This equation is equivalent to (5.7). Next we can condition the second term of the integral in (5.14) on Φ_{t-1} using the Theorem of Total Probability.

$$p(s_{t-1}, \Theta, \Phi_t \mid z^{t-1}, u^{t-1}, u_\phi^t, n^{t-1}) = \\ \int p(s_{t-1} \Theta, \Phi_t \mid \Phi_{t-1}, z^{t-1}, u^{t-1}, u_\phi^t, n^{t-1}) p(\Phi_{t-1} \mid z^{t-1}, u^{t-1}, u_\phi^t, n^{t-1}) d\Phi_{t-1} \tag{5.15}$$

The first term of (5.15) can be broken up using the definition of conditional probability.

$$p(s_{t-1}\Theta, \Phi_t \mid \Phi_{t-1}, z^{t-1}, u^{t-1}, u^t_\phi, n^{t-1}) = \tag{5.16}$$
$$p(\Phi_t \mid s_{t-1}, \Theta, \Phi_{t-1}, z^{t-1}, u^{t-1}, u^t_\phi, n^{t-1}) p(s_{t-1}, \Theta \mid \Phi_{t-1}, z^{t-1}, u^{t-1}, u^t_\phi, n^{t-1})$$

The first term of (5.16) simplifies to the dynamic object motion model.

$$p(s_{t-1}\Theta, \Phi_t \mid \Phi_{t-1}, z^{t-1}, u^{t-1}, u^t_\phi, n^{t-1}) = \tag{5.17}$$
$$\underbrace{p(\Phi_t \mid \Phi_{t-1}, u_{\phi,t})}_{\text{dynamic motion model}} \; p(s_{t-1}, \Theta \mid \Phi_{t-1}, z^{t-1}, u^{t-1}, u^t_\phi, n^{t-1})$$

Next, we substitute (5.17) back into (5.15), and combine terms using the definition of conditional probability.

$$p(s_{t-1}, \Theta, \Phi_t \mid z^{t-1}, u^{t-1}, u^t_\phi, n^{t-1}) = \tag{5.18}$$
$$\int p(\Phi_t \mid \Phi_{t-1}, u_{\phi,t}) p(s_{t-1}, \Theta, \Phi_{t-1} \mid z^{t-1}, u^{t-1}, u^t_\phi, n^{t-1}) d\Phi_{t-1}$$

Finally, the control $u_{\phi,t}$ can be dropped from the last term in (5.18) by the Markov property. The resulting equation is equivalent to (5.6).

5.1.2 Factoring the Dynamic SLAM Problem

FastSLAM's computational efficiency is a result of being able to factor the SLAM problem into independent estimation problems conditioned on the path of the robot. This factorization, initially presented in Chapter 3, is repeated for convenience:

$$p(s_t, \Theta \mid z^t, u^t, n^t) = p(s^t \mid z^t, u^t, n^t) \prod_{n=1}^{N} p(\theta_n \mid s^t, z^t, u^t, n^t) \tag{5.19}$$

The proof of this factorization, given in section 3.2.1, makes no assumptions about the motion of the features. It applies equally well to dynamic objects and static objects, as long as the motions of the different dynamic objects are independent. Replacing Θ with Θ and Φ_t, and adding the person controls u^t_ϕ, the factorization becomes:

$$p(s_t, \Theta, \Phi_t \mid z^t, u^t, u^t_\phi, n^t) = \tag{5.20}$$
$$p(s^t \mid z^t, u^t, n^t) \prod_{n=1}^{N_\theta} p(\theta_n \mid s^t, z^t, u^t, n^t) \prod_{n=1}^{N_\phi} p(\phi_n \mid s^t, z^t, u^t_\phi, n^t)$$

This posterior can also be implemented with a Rao-Blackwellized particle filter. With static landmarks, the landmark filters are updated exactly like they are in the original algorithm. For dynamic landmarks, an action update must be performed on the landmark filter at every time step, as well as a measurement update. This difference reflects the fact that the dynamic objects' positions become more uncertain as they move.

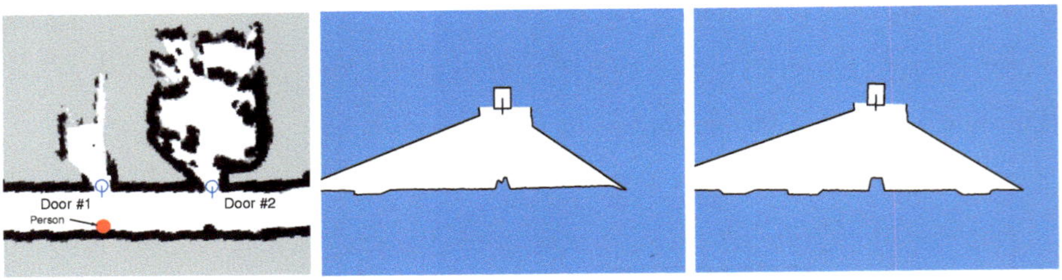

(a) A robot is in one of two doorways. (b) The robot's view from door #1 looking at a person (c) The robot's view from door #2 looking at a trash-bin

Fig. 5.1. Simultaneous Localization and People-Tracking.

5.2 Simultaneous Localization and People Tracking

Estimating whether each landmark is static or dynamic can be a daunting problem [94]. At this point, we will make the restrictive assumption that an accurate map of the static environment is known a priori. The problem now becomes tracking a set of dynamic features relative to a moving robot in a known map. In the experiments presented in this chapter, the dynamic features are people moving in the vicinity of a mobile robot. Localization and people tracking form a chicken-or-egg problem very similar to the SLAM problem [61]. Error in the pose of the robot correlates error in the estimated positions of the people. If the true pose of the robot is known, then the people can be tracked using independent filters. If the true positions of the people are known, then the position of the robot can be determined with maximum accuracy. If both are unknown, localization and people tracking become a joint estimation problem. Henceforth, we will refer to this estimation problem as Simultaneous Localization and People Tracking, or SLAP. The solution to the SLAP problem using a Rao-Blackwellized Particle Filter will be referred to as FastSLAP.

To illustrate the simultaneous nature of the SLAP problem, consider a robot operating in the map section shown in Figure 5.1(a). A person is in the hallway, standing in front of door #1, and the robot is in one of two locations specified in the map. If the robot is also in doorway #1 facing into the hallway, it sees the person and acquires the laser scan shown in Figure 5.1(b). If the robot is instead in door #2, the robot sees a trashcan in the hallway and acquires the laser scan shown in Figure 5.1(c). While the two scans look remarkably similar, they represent significantly different hypotheses. A localization algorithm that does not consider the first hypothesis will assume the person is the trashcan, and choose the wrong position for the robot. A people-tracking algorithm that does not consider the second hypothesis will track the trashcan as if it were a person.

5.2.1 Comparison with Prior Work

The majority of prior work in people tracking has been appearance-based methods. These algorithms attempt to detect the appearance of people in sensors, and track these features over time. Many examples of people tracking have used cameras as their primary sensor [30, 42], however laser range finders have also been used [81]. The accuracy of these approaches is limited primarily by the accuracy of the feature detection algorithm. In particular, dramatic variations in the appearance of a person due to illumination, viewing angle, and individual appearance, can make robust detection using vision an extraordinarily difficult problem.

Instead of detecting people by finding their appearance in a sensor, SLAP detects and tracks people using a map. If the position of the robot were known with certainty, all of the sensor readings corresponding to the static map could be subtracted away, leaving only the measurements of the people. Each person could be tracked with a separate low dimensional filter. If the position of the robot is uncertain, the person locations become correlated and have to be estimated jointly. Factoring the SLAP problem using the path of the robot allows us to track people using the map subtraction technique described above, even with significant uncertainty in the robot pose.

5.3 FastSLAP Implementation

Each particle in FastSLAP contains a robot pose $s_t^{[m]}$, and a set of $N_\phi^{[m]}$ person filters conditioned on the path of the particle. There are M_R particles in the particle filter.

$$S_t^{[m]} = \left\langle s_t^{[m]}, N_{\phi,t}^{[m]}, \Phi_{1,t}^{[m]}, \Phi_{2,t}^{[m]}, \ldots, \Phi_{N_{\phi^{[m]},t},t}^{[m]} \right\rangle \tag{5.21}$$

A variety of different kinds of filters can be used to implement the person filters $\Phi_{n,t}^{[m]}$. FastSLAM implements the landmark filters using EKFs. For the experiments shown in this chapter, we have chosen to implement the person filters as particle filters. A particle filter of particle filters has the same scaling properties as the original FastSLAM algorithm, however it can represent arbitrary posterior distributions (given enough particles).

The person filter $\Phi_{n,t}^{[m]}$ consists of M_P particles describing the position of the n-th person relative to the robot pose $s_t^{[m]}$.

$$\Phi_{n,t}^{[m]} = \{\phi_{n,t}^{[m][i]}\}_{i=1\ldots M_P} \tag{5.22}$$

The FastSLAP algorithm parallels the FastSLAM algorithm. A new robot pose is drawn for each particle S_{t-1} given the control u_t. Then an uninformative action u_ϕ is incorporated into all of the person filters. The true action of the person cannot be observed, so the action u_ϕ is merely a function of the

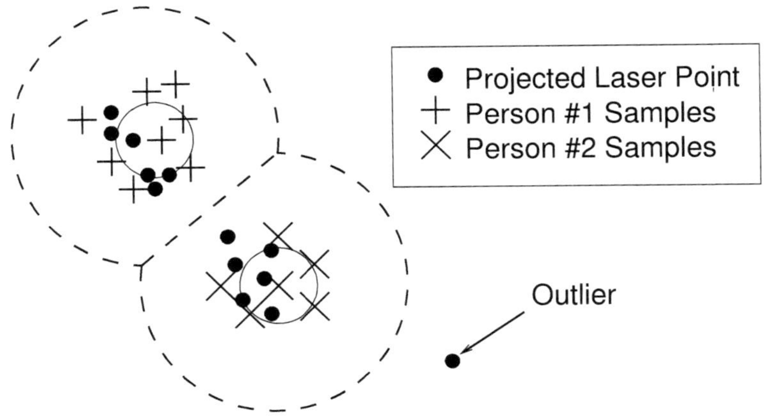

Fig. 5.2. Data Association in FastSLAP.

time between updates. Then, the observation z_t is assigned to one of $N_{\phi,t}^{[m]}+2$ different classes in each particle. It is assigned as either coming from the map or one of the $N_{\phi,t}^{[m]}$ existing person tracks, or used to create a new person track. If the observation is assigned to a person, the appropriate person filter is updated. Finally, the new robot particles are assigned weights, and a new set of particles S_t is drawn. The complete algorithm is described in Table 5.1.

The details of FastSLAP are largely similar FastSLAM. The remainder of this section will be devoted to describing the differences between the two algorithms.

5.3.1 Scan-Based Data Association

The standard formulation of the SLAM problem assumes that the robot makes discrete observations of features in the environment. FastSLAP was developed for a robot with a 2-D laser range finder. This sensor produces a scan of 180 range readings covering a wide field-of-view. Instead of producing observations of individual objects, the range finder images its entire environment. In order to use the scans for localization and people tracking, each range reading must be assigned to one of $N_{\phi,t}^{[m]}+2$ classes for each particle.

If every sensor reading contributed evidence into every person filter, all $N_{\phi,t}^{[m]}$ filters would track the one most probable person. The associations between observations and filters can be a *hard* assignment, in which each observation is attributed to only one filter, or a *soft* assignment in which each observation can be assigned in part to multiple filters.

When two different filters are far apart, the difference between the hard and soft assignment strategy is minimal. However, when filters are close, soft assignment tends to lump the two filters together. Both filters accept approximately fifty percent assignment of all of the sensor readings originally

Table 5.1. The FastSLAP Algorithm.

Algorithm FastSLAP(S_{t-1}, z_t, u_t)
$S_t = S_{aux} = \emptyset$
for $m = 1$ **to** M_R // loop over robot particles
  retrieve m-th particle $\left\langle s_{t-1}^{[m]}, N_{t-1}^{[m]}, \Phi_{1,t-1}^{[m]}, \ldots, \Phi_{N_{t-1}^{[m]},t-1}^{[m]} \right\rangle$ from S_{t-1}
  draw $s_t^{[m]} \sim p(s_t \mid s_{t-1}^{[m]}, u_t)$
  for $n = 1$ **to** $N_{t-1}^{[m]}$ // loop over person filters
    for $i = 1$ **to** M_P
      draw $\phi_{n,t}^{[m][i]} \sim p(\phi_t \mid \phi_{n,t-1}^{[m][i]}, u_\phi)$ // draw new people particles from person motion model
    end for
  end for
  determine data association n_t **for** z_t
  if $n_t = N_{t-1}^{[m]} + 1$ // new person?
    $N_t^{[m]} = N_{t-1}^{[m]} + 1$
    for $i = 1$ **to** M_P
      $\phi_{n_t,t}^{[m][i]} \sim \mathcal{N}(\phi_t; g^{-1}(s_t^{[m]}, z_t), \Sigma_{\Phi,0})$ // initialize person particles
    end for
    $\Phi_{n_t,t}^{[m]} = \left\{ \phi_{n_t,t}^{[m][i]} \right\} |_{i=1\ldots M_P}$
  end if
  if $n_t = n_\Theta$ // observation of the map?
    $w_t^{[m]} = p(z_t \mid s_t^{[m]}, \Theta, n_\Theta)$
  else // observation of a person?
    $\Phi_{n_t,aux}^{[m]} = \emptyset$
    for $i = 1$ **to** M_P
      $w_t^{[m][i]} = p(z_t \mid s_t^{[m]}, \phi_{n_t,t}^{[m][i]}, n_t)$
      add $\left\langle \phi_{n_t,t}^{[m][i]}, w_t^{[m][i]} \right\rangle$ **to** $\Phi_{n_t,aux}^{[m]}$
    end for
    $w_t^{[m]} = \Sigma_{i=1}^{M_P} w_t^{[m][i]}$
    draw M_P particles from $\Phi_{n_t,aux}^{[m]}$ with prob. $\propto w_t^{[m][i]}$, add to $\Phi_{n_t,t}^{[m]}$
    // resample person filter
  end if
  for $i = 1$ **to** $N_t^{[m]}$
    if $i \neq n_t$ // do not resample
      $\Phi_{n,t}^{[m]} = \left\{ \phi_{n,t}^{[m][i]} \right\} |_{i=1\ldots M_P}$ unobserved people filters
      delete $\Phi_{n,t}^{[m]}$ if it hasn't been observed **for** many iterations
      // delete unused person filters
    end if
  end for
  add $\left\langle s_t^{[m]}, N_t^{[m]}, \Phi_{1,t}^{[m]}, \ldots, \Phi_{N_t^{[m]}}^{[m]}, w_t^{[m]} \right\rangle$ **to** S_{aux}
end for
draw M_R particles from S_{aux} with prob. $\propto w_t^{[m]}$, add to S_t // resample robot filter
return S_t

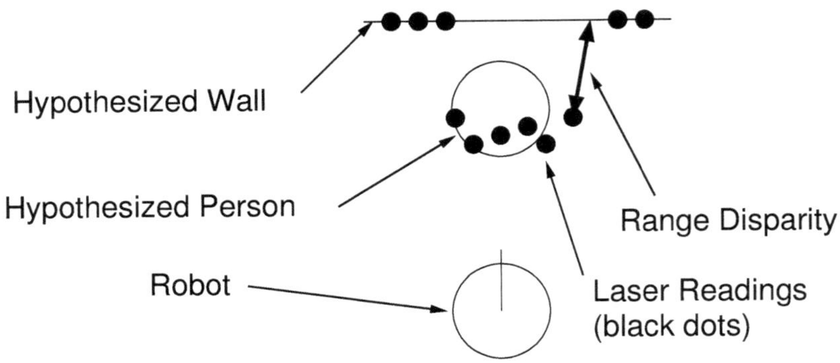

Fig. 5.3. Effect of a non-smooth measurement model.

associated with the two filters. Hard assignment, on the other hand, continues to provide good discrimination between the two filters even as the filters become very close.

The actual assignment of each laser point is determined using a nearest neighbor rule. Each laser point is assigned to the nearest person filter using Mahalanobis distance, if it is within a certain maximum distance. If an observation is beyond the maximum distance, it is considered an outlier and ignored, or it is used to create a new person filter.

5.3.2 Measurement Model

Clearly, the form of the measurement model depends on the sensor being used by the robot. However, the exact form of the measurement model can have important consequences for the performance of the filter. In the experiments described in this chapter, a mobile robot was equipped with a 2D laser range finder. This range finder is able to determine the distance to objects in the vicinity of the robot over a 180 degree field-of-view.

The measurement model $p(z_t \mid s_t, \Theta, \Phi_t, n_t)$ characterizes the probability of receiving a sensor reading given a particular joint state of the robot and a person. In other words, the model compares what the robot actually senses with what it should "expect" to sense given the hypothesized state. Typically this model is based upon the physics of the real-world sensor being used and its interaction with its environment. A person appearing in a laser range finger might be modeled crudely as a cylinder. The measurement model can be computed by comparing the actual laser scan with a laser scan simulated according to the hypothesized states of the robot and person in the map.

Unfortunately, small differences between the true and hypothesized states of the world can result in large differences in the probabilities of the resulting scans. Consider the situation depicted in Figure 5.3. A laser measurement, expected to pass by the person, actually hits the person. This disparity in range causes the individual reading and thus the entire scan to receive a very low probability. A larger number of particles will be necessary to estimate the state of the person using this measurement model. In general, lack of smoothness in the measurement model will require a higher number of to be used [55].

The sensor model can be made much smoother by calculating disparities in x, y space, rather than disparities in range. To calculate the probability of a given state, the laser readings are first projected into the world according to the hypothesized pose of the robot. The probability of each laser point is then computed based on the Euclidean distance between that point and the closest object in the world, be it a hypothesized person or an occupied map cell. Using this sensor model, the mismatched point in Figure 5.3 would receive a high probability because it is close to a hypothesized person. The construction of a smooth measurement model significantly reduces the number of samples necessary to a achieve a particular tracking accuracy.

5.3.3 Motion Model

The motion models $p(s_t \mid s_{t-1}, u_t)$ and $p(\Phi_t \mid \Phi_{t-1}, u_\phi)$ predict the movement over time of the robot and of people, respectively. The robot's motion model is well understood and was taken directly from [89]. However, no information analogous to odometry is available to describe the motion of the people. Instead, we will use a Brownian motion model for the human motion that attempts to predict all motions of people over a short time interval. This model assumes that people move according to a velocity that is constantly being perturbed randomly. The variance of the perturbation acts like momentum. The lower the variation, the more likely the person is to move in the direction of past motion.

$$v_{t+1} = N(v_t, \sigma_v^2) \tag{5.23}$$

$$s_t = s_t + v_{t+1} \Delta t \tag{5.24}$$

5.3.4 Model Selection

The person filters operate on the assumption that the true number of people N_t is known. In practice, determining N_t can be a difficult proposition. People are constantly occluding each other in the robot's field-of-view, especially as they walk very close to the robot. If a person is temporarily occluded, N_t should not change. However, people also move in and out of the robot's field-of-view in a permanent way, going around corners and walking into offices. If a person disappears for an extended period of time, the SLAP algorithm should respond by changing the value of N_t.

One approach to determining N_t is to create a prior distribution over "typical" values of N, and choose the value of N_t at every time step that corresponds with the Minimum Description Length (MDL) hypothesis [78, 41]. This approach will add a new person filter only if it results in a significant gain in the overall probability of the model. However, this approach requires that multiple instances of every robot particle be run with different values of N_t. As a result, significant computation is spent on filters that do not represent the true state of the world.

The MDL approach can be approximated in a practical manner by examining the data associations of the observations. A cluster of sensor readings that are not associated with any person filter probably indicates that N_t is too small. A filter that has no observations assigned to it for an extended period of time indicates that N_t is too large. Experimental results in this chapter illustrate that heuristics based on the associations of the laser readings can be used to determine N_t with high accuracy at low computational cost.

5.4 Experimental Results

5.4.1 Tracking and Model Selection Accuracy

FastSLAP was tested on a robot with a laser range finder operating in a typical office environment. Figure 5.4(a) shows a typical laser scan given to the algorithm. The scanner is approximately 12 inches off the ground, so it detects the individual legs of people walking near the robot. Figure 5.4(b) shows the state of the SLAP filter after incorporating the laser scan. The people filters drawn correspond to the most likely robot particle. Both people within range of the robot are being tracked successfully.

The accuracy of localization and people-tracking were evaluated based on hand-labeled ground truth data captured from a second, fixed laser range finder. The standard deviation of the position error of the robot was approximately 6 cm, and less than 5 cm for the positions of the people. The mean errors of the robot and the people positions were both less than 3 cm.

The accuracy of model selection was tested on a data set approximately six minutes long. Over the course of the run, 31 people passed within the sensor range of the moving robot. At any given time, up to four people were simultaneously visible to the robot. Of those 31 people, only 3 were not tracked correctly. In one instance, two people entered the robot's view in close proximity, and walked very closely to each other. In that situation, the two people were tracked as a single entity.

Model selection was also tested in a more difficult environment, in which the map was not up-to-date. In this run, the robot encountered 11 different people, up to 5 at a time. All 11 people in this environment were tracked correctly. However, the algorithm also tracked an additional 4 objects. These tracks corresponded with inanimate objects that were not in the map, including a recycling bin, a chair, and a closed door which was open when the map was made.

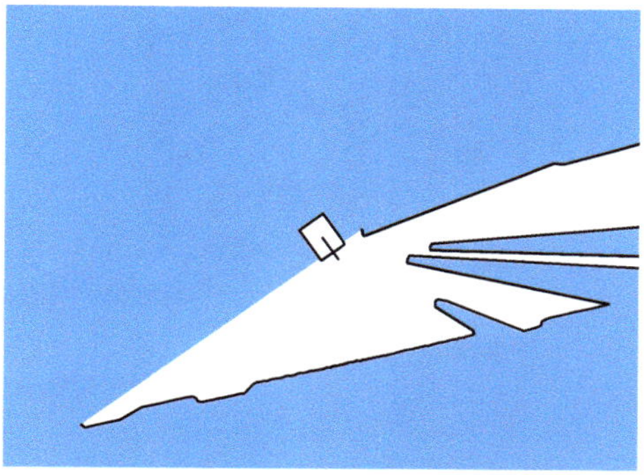

(a) Laser scan showing two people near the robot

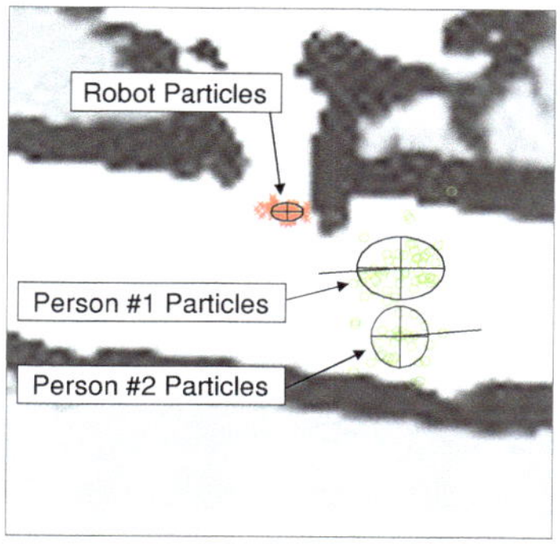

(b) Output of the SLAP filter showing the estimated position of the robot and the people in the map

Fig. 5.4. Typical input and output of FastSLAP.

5.4.2 Global Uncertainty

Figure 5.4 and Table 5.2 illustrate the performance of the SLAP filter in situations in which the pose of the robot is relatively well known. In these situations, people trackers that ignore the pose uncertainty of the robot would also work well. The real power of this approach is demonstrated in situations in which there is significant uncertainty in the robot pose estimate. This commonly occurs during global localization, when a robot is initialized with no prior information about its position or orientation relative to the map.

Figure 5.5 shows the output of the SLAP filter during global localization with a single person in the robot's field of view. Figure 5.5(a) shows the filter just after initialization, with robot and person particles scattered all over the map. After the robot moves a few meters as shown in Figure 5.5(b), two modes develop in the robot and people distributions. The two modes correspond with the robot having started from either one of two different doorways in a relatively uniform hallway. Even though there is significant uncertainty in the position of the robot, the person is still being tracked correctly relative to the two modes. This is evidenced by the two clusters of people particles moving ahead of the two robot modes. At the robot moves further down the hallway, sensor evidence eventually disambiguates between the two primary hypotheses and the filter converges on the true state of the world, shown in Figure 5.5(c).

5.4.3 Intelligent Following Behavior

A following behavior was implemented using the output of FastSLAP. Independent control loops governing the rotational and translational velocity of

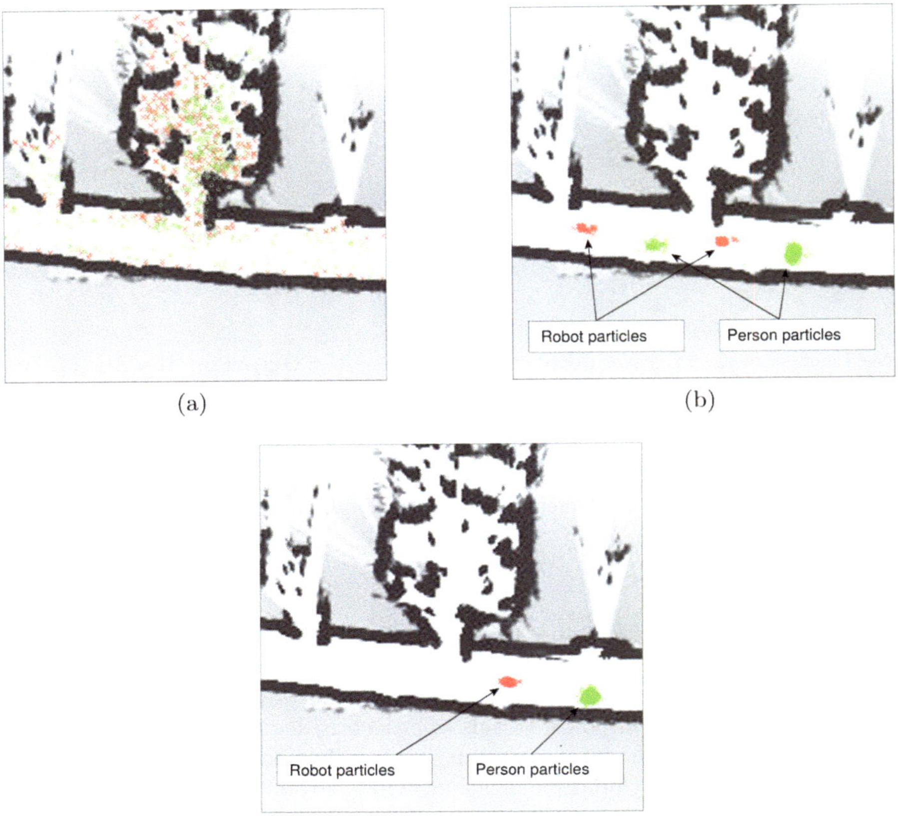

(a) (b)

(c)

Fig. 5.5. Evolution of FastSLAP filter from global uncertainty to successful localization and tracking.

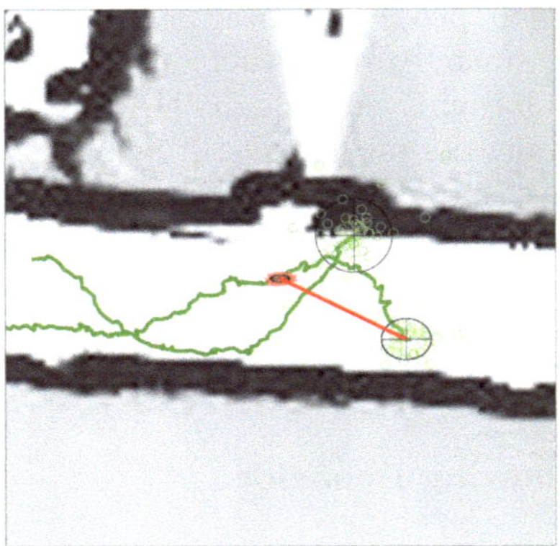

Fig. 5.6. Intelligent following behavior based on the output of FastSLAP.

Table 5.2. Performance of FastSLAP.

Tracking Accuracy

Robot position – mean error	2.5 cm
Robot position – standard deviation of error	5.7 cm
People positions – mean error	1.5 cm
People positions – standard deviation of error	4.2 cm

Model Selection

True number of people (cumulative)	31
Model selection errors	3
Model selection accuracy	90%

the robot were based on the relative range and bearing to the subject being followed. A snapshot of the behavior in operation is shown in Figure 5.6. The robot was instructed to follow one of two people within range of the robot. A thick line is drawn between the mean robot position and position of the person being followed. The robot successfully followed the person down the hallway, as the second person repeatedly walked between the subject and the robot. The robustness of the people-tracker to occlusion allows a simple control loop to follow a person reliably, even in crowded environments.

5.5 Summary

This chapter presented an extension of the SLAM problem to worlds with static and dynamic features. By assuming that a map of the static world is available a priori, a map-based people tracking algorithm was developed that can track the positions of people even with global uncertainty over the pose of the robot. The FastSLAP algorithm is efficient enough to track large numbers of people while still being robust to uncertainty in the pose of the robot. Output of FastSLAP was used to create a intelligent following behavior for a mobile robot that is robust to repeated occlusion.

6

Conclusions

6.1 Conclusion

This book described FastSLAM, a scalable approach to the Simultaneous Localization and Mapping Problem with unknown data association. FastSLAM differs from existing approaches in that it exploits sparsity in the dependencies between data and the state variables *over time* to factor the SLAM problem into a set of low-dimensional problems. FastSLAM samples over the robot's path and data associations, and computes independent landmark estimates conditioned on each particle.

FastSLAM has several advantages over EKF-based approaches to the SLAM problem.

- **Logarithmic Complexity.** Each particle in the FastSLAM particle filter is organized as a binary tree of landmark filters. Organizing the landmarks in trees allows subtrees to be shared between multiple particles. If only a small number of landmarks are observed at any time, then this overlap will be substantial. As a result, observations can be incorporated into the filter in time logarithmic with the total number of landmarks in the map. The log(N) algorithm has been applied to maps with up to 1,000,000 landmarks. Sharing subtrees also results in a substantial reduction in memory usage over the linear time FastSLAM algorithm.
- **Multi-Hypothesis Data Association.** Each FastSLAM particle represents a single hypothesized path of the robot. Consequently, different data association hypotheses can be applied to each particle. Multi-hypothesis data association has several advantages over single-hypothesis data association. First, it factors robot pose uncertainty out of the data association problem. Unlike the EKF, where the landmark uncertainty is a function of both motion error and measurement error, landmark filters in FastSLAM are conditioned on a specific path of the robot. Without the effects of motion error, the identities of landmarks are more easily determined.

The second benefit of maintaining multiple hypotheses is the capability to delay data association decisions until future observations are acquired. The correct data association is not always the most probable choice given a single observation. Unlike the EKF, FastSLAM can throw out data associations made in the past if future observations show that they are incorrect. Experimental results demonstrated that multi-hypothesis data association allows FastSLAM to build maps successfully in simulated and real world environments with substantial data association ambiguity.
- **Convergence without the full covariance matrix.** This book presented a convergence proof for FastSLAM 2.0 in linear-Gaussian worlds. This proof has little practical importance because real-world SLAM problems are rarely linear. However, it demonstrates that maintaining the full covariance matrix of the Kalman Filter is not strictly necessary to insure convergence. In fact, FastSLAM with a single particle, a special case of the algorithm that maintains no cross-correlations between landmarks, also converges. To our knowledge, this is the first constant time algorithm for which convergence has been shown. In practice the constant time version of FastSLAM will converge much more slowly than the EKF, however this approach demonstrates that algorithms need not be conservative in order to converge.

6.1.1 Limitations of FastSLAM

FastSLAM maintains samples over the space of possible robot trajectories. The resampling step of the algorithm throws away low probability samples while duplicating high probability samples multiple times. This process is necessary to bring the particles into agreement with the latest observations, however it gradually eliminates diversity in the trajectories represented by the samples. As a result, all FastSLAM particles will share a common trajectory at some point in the past. New observations cannot affect landmarks that last observed prior to this point of common history. The resampling step of FastSLAM can be thought of as a process for throwing correlation information away smoothly over time. Throwing away correlation data creates two different problems, both of which are symptoms of a larger issue.

- **Underestimated Covariance.** Ignoring correlation information will inevitably cause FastSLAM to underestimate the covariance of landmarks in the map. This will result in lower map accuracy given known data association, and it can complicate data association decisions if the correspondences are unknown. The thresholds for data association in FastSLAM must be more "accepting" to account for underestimated landmark uncertainty. This may make the process for adding a new landmark more difficult as well.
- **Loop Closing.** Loop closing is generally agreed to be one of the most difficult cases of the SLAM problem. Initially, FastSLAM's multi-hypothesis

approach seems well suited to solving the data association problem induced by the large pose uncertainty at the end of the loop. However, throwing away correlation data reduces the effect of observations on landmarks around the loop. In order to match the performance of the EKF, the number of FastSLAM particles must grow as a function of the size of the loop being closed. At some point, the number of parameters in the particle filter will exceed that of the EKF to close a certain sized loop.

The problems of underestimated covariance and loop closing are both a result of the relationship between correlation information and convergence speed in FastSLAM. FastSLAM's computational advantage over the EKF is a consequence of throwing away correlation information. However, throwing away correlation slows down the rate at which observations can affect the positions of other landmarks, slowing the convergence of the map. EKF submap approaches suggest that a great deal of correlation information can be thrown away without seriously compromising convergence speed. The disadvantage of throwing away correlation data over time, as opposed to discarding correlation information spatially, is that the performance of the algorithm becomes more dependent on the specific control policy executed by the robot.

6.2 Future Work

There are a number of open research issues related to FastSLAM.

- **Convergence rate as a function of particles.** Experimental results in this book demonstrate that the number of particles necessary to close a loop with FastSLAM increases as a function of loop size. Currently, it is unknown whether this function is polynomial or exponential. Determining the relationship between the number of particles and convergence speed will give us a better understanding of the limitations of FastSLAM.
- **Multi-robot mapping.** Sampling over robot pose enables a special case of the multi-robot SLAM problem to be implemented easily. Specifically, if a robot is placed in an uncertain map created by another robot, FastSLAM can be used to add information to the map, even if there is global uncertainty over the pose of the new robot. Sampling over robot pose eliminates robot pose uncertainty from the data association problem, which is substantial during global localization.
- **Optimal control problem.** The accuracy of the map generated FastSLAM is a function not only of motion and measurement error, but also the control policy of the robot. A control policy that generates smaller loops will clearly perform better than a policy that generates large loops. Determining the optimal action for the robot to take given the current map is an extremely challenging problem in environments with large numbers of landmarks.

References

1. T. Bailey. *Mobile Robot Localisation and Mapping in Extensive Outdoor Environments.* PhD thesis, University of Sydney, 2002.
2. D. Ballard and C. Brown. *Computer Vision*, chapter 4. Prentice-Hall, 1982.
3. Y. Bar-Shalom and T. Fortmann. *Tracking and Data Association.* Academic Press, Inc., 1988.
4. P. Besl and N. McKay. A method for reigstration of 3d shapes. *PAMI*, 14(2):239–256, 1992.
5. M. Bosse, P. Newman, J. Leonard, and S. Teller. An atlas framework for scalable mapping. Technical report, Massachusetts Institute of Technology, 2002.
6. R. Chellappa and A. Jain, editors. *Markov Random Fields: Theory and Applications.* Academic Press, 1993.
7. K. Chong and L. Kleeman. Feature based mapping in real, large scale environments using an ultrasonic array. *International Journal of Robotics Research*, 18(1):3–19, 1999.
8. H. Christensen, editor. *Lecture Notes: SLAM Summer School.* Unpublished, 2002. http://www.cas.kth.se/SLAM/toc.html.
9. W. Cochran. *Sampling Techniques, Third Edition.* John Wiley and Sons, 1977.
10. A. J. Davison. *Mobile Robot Navigation Using Active Vision.* PhD thesis, University of Oxford, 1998.
11. N. de Freitas, M. Niranjan, A. Gee, and A. Doucet. Sequential monte carlo methods to train neural networks. *Neural Computation*, 12(4), 2000.
12. M. Deans. *Bearings-Only Localization and Mapping.* PhD thesis, Carnegie Mellon University, 2002.
13. M. Deans and M. Hebert. Experimental comparison of techniques for localization and mapping using a bearing-only sensor. In *Proceedings of the International Symposium on Experimental Robotics (ISER)*, 2002.
14. F. Dellaert. *Monte Carlo EM for Data-Association and its Application in Computer Vision.* PhD thesis, Carnegie Mellon University, 2001.
15. F. Dellaert. Square root SAM. In S. Thrun, G. Sukhatme, S. Schaal, and O. Brock, editors, *Proceedings of the Robotics Science and Systems Conference*, Cambridge, MA, 2005.
16. F. Dellaert, D. Fox, W. Burgard, and S. Thrun. Monte carlo localization for mobile robots. In *Proceedings of the IEEE International Conference on Robotics and Automation*, 1999.

17. A. Dempster, A. Laird, and D. Rubin. Maximum likelihood from incomplete data via the em algorithm. *Journal of the Royal Statistical Society, Series B*, 39(1):1–38, 1977.
18. G. Dissanayake, P. Newman, H.F. Durrant-Whyte, S. Clark, and M. Csobra. *Lecture Notes in Control and Information Sciences, Experimental Robotics VI*. Springer-Verlag, 2000.
19. M. Dissanayake, P. Newman, S. Clark, H. Durrant-Whyte, and M. Csorba. A solution to the simultaneous localization and mapping (slam) problem. *IEEE Transactions on Robotics and Automation*, 2001.
20. A Doucet. On sequential simulation-based methods for Bayesian filtering. Technical Report CUED/F-INFENG/TR 310, Cambridge University, Department of Engineering, Cambridge, UK, 1998.
21. A. Doucet, N. de Freitas, and N. Gordon, editors. *Sequential Monte Carlo Methods in Practice*. Springer-Verlag, 2001.
22. A. Doucet, N. de Freitas, K. Murphy, and S. Russell. Rao-blackwellised particle filtering for dynamic bayesian networks. In *Proceedings of the Sixteenth Conference on Uncertainty in Artificial Intelligence*, pages 176–183, Stanford, 2000.
23. A. Doucet, N. de Freitas, K. Murphy, and S. Russell. Rao-blackwellised particle filters for dynamic bayes nets. In *Proceedings of Conference on Uncertainty in Artificial Intelligence*, 2000.
24. T. Duckett, S. Marsland, and J. Shapiro. Fast, on-line learning of globally consistent maps. *Autonomous Robots*, 12(3):287 – 300, 2002.
25. A. Elfes. Sonar-based real-world mapping and navigation. *IEEE Journal of Robotics and Automation*, RA-3(3):249–265, June 1987.
26. A. Eliazar and R. Parr. DP-SLAM: Fast, robust simultaneous localization and mapping without predetermined landmarks. In *Proceedings of the Sixteenth International Joint Conference on Artificial Intelligence (IJCAI)*, Acapulco, Mexico, 2003. IJCAI.
27. M. Fischler and R. Bolles. Random sample consensus: A paradigm for model fitting with applications to image analysis and automated cartography. *CACM*, 24(6):381–395, June 1981.
28. J. Folkesson and H. I. Christensen. Graphical SLAM: A self-correcting map. In *Proceedings of the IEEE International Conference on Robotics and Automation (ICRA)*, New Orleans, USA, 2004.
29. U. Frese and G. Hirzinger. Simultaneous localization and mapping—a discussion. In *Proceedings of the IJCAI Workshop on Reasoning with Uncertainty in Robotics*, pages 17–26, Seattle, WA, 2001.
30. D. M. Gavrila. The visual analysis of human movement: A survey. *Computer Vision and Image Understanding: CVIU*, 73(1):82–98, 1999.
31. A. Gelb. *Applied Optimal Estimation*. MIT Press, 1989.
32. T. Gibb. Quecreek commission says better maps are a must. Pittsburgh Post Gazette, November 26, 2002.
33. M. Golfarelli, D. Maio, and S. Rizzi. Elastic correction of dead-reckoning errors in map building. In *Proceedings of the IEEE/RSJ International Conference on Intelligent Robots and Systems (IROS)*, pages 905–911, Victoria, Canada, 1998.

34. M. P. Golombek, R. A. Cook, T. Economou, W. M. Folkner, A. F. Haldemann, P. H. Kallemeyn, J. M. Knudsen, R. M. Manning, H. J. Moore, T. J. Parker, R. Rieder, J. T. Schofield, P. H. Smith, and R. M. Vaughan. Overview of the mars pathfinder mission and assessment of landing site predictions. *Science*, 278(5344):1743–1748, December 1997.
35. W. Grimson and T. Lozano-Perez. Localizing overlapping parts by searching the interpretation tree. *IEEE Transactions on Pattern Analysis and Machine Intelligence*, 9(4):469–482, 1987.
36. J. Guivant and E. Nebot. Optimization of the simultaneous localization and map building algorithm for real time implementation. *IEEE Transactions on Robotics and Automation*, 17(3):242–257, 2001.
37. J. Guivant, E. Nebot, and S. Baiker. Autonomous navigation and map building using laser range sensors in outdoor applications. *Journal of Robotics Systems*, 17(10):565–583, 2000.
38. J.-S. Gutmann and B. Nebel. Navigation mobiler roboter mit laserscans. In *Autonome Mobile Systeme*. Springer Verlag, Berlin, 1997. In German.
39. D. Haehnel, D. Schultz, and W. Burgard. Map building with mobile robots in populated environments. In *Proceedings of the International Conference on Intelligent Robots and Systems (IROS)*, 2002.
40. D. Hähnel, D. Fox, W. Burgard, and S. Thrun. A highly efficient FastSLAM algorithm for generating cyclic maps of large-scale environments from raw laser range measurements. In *Proceedings of the Conference on Intelligent Robots and Systems (IROS)*, 2003.
41. M. Hansen and B. Yu. Model selection and the principle of minimum description length. *JASA*, 96(454):746–774, 2001.
42. I. Haritaoglu, D. Harwood, and L. Davis. A real time system for detecting and tracking people. In *Proceedings of the International Conference on Automatic Face and Gesture Recognition*, Nara, Japan, 1998.
43. R. Hartley. Euclidean reconstruction from uncalibrated views. In J. Mundy, A. Zisserman, and D. Forsyth, editors, *Applications of Invariance in Computer Vision*, pages 237–256. Springer-Verlag, 1994.
44. M. Isard and A. Blake. Condensation – conditional density propogation for visual tracking. *International Journal of Computer Vision*, 29(1):5–28, 1998.
45. S. Julier and J. Uhlmann. Building a million beacon map. In *SPIE Sensor Fusion*, 2001.
46. R. Kalman. A new approach to linear filtering and prediction problems. *Transactions of the ASME–Journal of Basic Engineering*, 82(Series D):35–45, 1960.
47. J. Knight, A. Davison, and I. Reid. Constant time slam using postponement. In *Proceedings of IEEE International Conference on Intelligent Robots and Systems (IROS)*, 2001.
48. K. Konolige. Large-scale map-making. In *Proceedings of the AAAI National Conference on Artificial Intelligence*, pages 457–463, San Jose, CA, 2004. AAAI.
49. J. Leonard and H. Feder. A computationally efficient method for large-scale concurrent mapping and localization. In J. Hollerbach and D. Koditschek, editors, *Proceedings of the Ninth International Symposium on Robotics Research*, London, 2000. Springer-Verlag.
50. J. Leonard, R. Rickoski, P. Newman, and M. Bosse. Mapping partially observable features from multiple uncertain vantage points. *International Journal of Robotics Research*, 2002. to appear.

51. J. Liu and R. Chen. Sequential monte carlo methods for dynamic systems. *Journal of the American Statistical Association*, 93:1032–1044, 1998.
52. D. Lomet and B. Salzberg. The hb-tree: A multiattribute indexing method. *ACM Transactions on Database Systems*, 15(4):625–658, 1990.
53. F. Lu and E. Milios. Globally consistent range scan alignment for environment mapping. *Autonomous Robots*, 4:333–349, 1997.
54. F. Lu and E. Milios. Robot pose estimation in unknown environments by matching 2d range scans. *Journal of Intelligent and Robotic Systems*, 18:249–275, 1998.
55. D. MacKay. *Learning in Graphical Models*, chapter Introduction to Monte Carlo Methods, pages 175–204. Kluwer Academic Publishers, 1998.
56. W. Madow. On the theory of systematic sampling, ii. *Annals of Mathematical Statistics*, 20:333–354, 1949.
57. M. Montemerlo and S. Thrun. Simultaneous localization and mapping with unknown data association using fastslam. In *ICRA*, 2003.
58. M. Montemerlo and S. Thrun. Large-scale robotic 3-d mapping of urban structures. In *Proceedings of the International Symposium on Experimental Robotics (ISER)*, Singapore, 2004. Springer Tracts in Advanced Robotics (STAR).
59. M. Montemerlo, S. Thrun, D. Koller, and B. Wegbreit. Fastslam: A factored solution to the simultaneous localization and mapping problem. In *AAAI*, 2002.
60. M. Montemerlo, S. Thrun, D. Koller, and B. Wegbreit. Fastslam 2.0: An improved particle filtering algorithm for simultaneous localization and mapping that provably converges. In *IJCAI*, 2003.
61. M. Montemerlo, W. Whittaker, and S. Thrun. Conditional particle filters for simultaneous mobile robot localization and people tracking. In *Proceedings of the International Conference on Robotics and Automation (ICRA)*, 2002.
62. A. Moore. An introductory tutorial on kd-trees. Technical Report No. 209, University of Cambridge, 1991.
63. H. P. Moravec. Sensor fusion in certainty grids for mobile robots. *AI Magazine*, 9(2):61–74, 1988.
64. P. Moutarlier and R. Chatila. An experimental system for incremental environment modeling by an autonomous mobile robot. In *1st International Symposium on Experimental Robotics*, June 1989.
65. P. Moutarlier and R. Chatila. Stochastic multisensory data fusion for mobile robot location and environment modeling. In *5th International Symposium on Robotics Research*, Tokyo, 1989.
66. K. Murphy. Bayesian map learning in dynamic environments. In *Advances in Neural Information Processing Systems (NIPS)*. MIT Press, 1999.
67. E. Nebot, F. Masson, J. Guivant, and H. Durrant-Whyte. Robust simultaneous localization and mapping for very large outdoor environments. In *Proceedings of the 8th International Symposium on Experimental Robotics (ISER)*, 2002.
68. J. Neira and J. Tardos. Data association in stochastic mapping using the joint compatibility test. *IEEE Transactions on Robotics and Automation*, 17(6):890–897, 2001.
69. P. Newman. *On the Structure and Solution of the Simultaneous Localisation and Map Building Problem*. PhD thesis, University of Sydney, March 1999.

70. P. Newman, J. Leonard, J. Neira, and J. Tardos. Eplore and return: Experimental validation of real-time concurrent mapping and localization. In *Proceedings of the IEEE International Conference on Robotics and Automation (ICRA)*, 2002.
71. J. Niento, J. Guivant, E. Nebot, and S. Thrun. Real time data association in fastslam. In *Proceedings of the IEEE International Conference on Robotics and Automation*, 2003.
72. Notes: ICRA Workshop on Concurrent Mapping and Localization for Autonomous Mobile Robots, 2002.
73. M. A. Paskin. Thin junction trees filters for simultaneous localization and mapping. Technical Report UCB/CSD-02-1198, UC Berkeley, 2002.
74. J. Pearl. *Probabilistic Reasoning in Intelligent Systems: Networks of Plausible Inference.* Morgan Kaufmann, 1988.
75. M. Pitt and N. Shephard. Filtering via simulation: auxiliary particle filter. *Journal of the American Statistical Association*, 94:590–599, 1999.
76. O. Procopiuc, P. Agarwal, L Arge, and J. Vitter. Bkd-tree: A dynamic scalable kd-tree. Submitted for publication, 2002.
77. D. Reid. An algorithm for tracking multiple targets. *IEEE Transactions on Automatic Control*, 24(6):843–854, 1979.
78. J. Rissanen. On optimal number of features in classification. Technical Report RJ 6471, IBM Research Division, Yorktown Heights, N.Y., 1988.
79. D. B. Rubin. Using the SIR algorithm to simulate posterior distributions. In M. Bernardo, K. DeGroot, D. Lindley, and A. Smith, editors, *Bayesian Statistics 3*. Oxford University Press, 1988.
80. S. Russell and P. Norvig. *Artificial Intelligence: A Modern Approach.* Prentice-Hall, 1995.
81. D. Schulz, W. Burgard, D. Fox, and A. Cremers. Tracking multiple moving targets with a mobile robot using particles filters and statistical data association. In *Proceedings of the IEEE International Conference on Robotics and Automation*, Seoul, Korea, 2001.
82. R. C. Smith and P. Cheeseman. On the representation and estimation of spatial uncertainty. *International Journal of Robotics Research*, 5(4):56–68, 1986.
83. T. Stepelton. Personal communications.
84. J. Tardos, J. Niera, P. Newman, and J. Leonard. Robust mapping and localization in indoor environments using sonar data. *International Journal of Robotics Research*, 2002.
85. C. Thorpe and H. Durrant-Whyte. Field robots. In *Proceedings of the 10th International Symposium of Robotics Research (ISRR'01)*, Lorne, Australia, 2001.
86. S. Thrun. Robotic mapping: A survey. In G. Lakemeyer and B. Nebel, editors, *Exploring Artificial Intelligence in the New Millenium.* Morgan Kauffmann, 2002. to appear.
87. S. Thrun, W. Burgard, and D. Fox. *Probabilistic Robotics.* MIT Press, Cambridge, MA, 2005.
88. S. Thrun, D. Ferguson, D. Haehnel, M. Montemerlo, and W. Burgard R. Triebel. A system for volumetric robotic mapping of abandoned mines. In *Proceedings of the IEEE International Conference on Robotics and Automation (ICRA'03)*, 2003. to appear.

89. S. Thrun, D. Fox, and W. Burgard. Monte carlo localization with mixture proposal distribution. In *Proceedings of AAAI National Conference on Artificial Intelligence*, Austin, Texas, 2000.
90. S. Thrun, D. Koller, Z. Ghahramani, H. Durrant-Whyte, and A. Y. Ng. Simultaneous mapping and localization with sparse extended information filters. In *Proceedings of WAFR*, 2002.
91. C. Tomasi and T. Kanade. Shape and motion from image streams: A factorization method. Technical Report CMU-CS-92-104, Carnegie Mellon University, 1992.
92. R. van der Merwe, N. de Freitas, A. Doucet, and E. Wan. The unscented particle filter. In *Proceedings of NIPS*, 2001.
93. C.-C. Wang, C. Thorpe, and S. Thrun. Online simultaneous localization and mapping with detection and tracking of moving objects: Theory and results from a ground vehicle in crowded urban areas. In *Proceedings of the IEEE International Conference on Robotics and Automation (ICRA)*, 2003.
94. C.-C. Wang, C. Thorpe, and S. Thrun. Online simultaneous localization and mapping with detection and tracking of moving objects: Theory and results from a ground vehicle in crowded urban areas. In *Proceedings of the IEEE International Conference on Robotics and Automation (ICRA)*, 2003.
95. G. Welch and G. Bishop. An introduction to the kalman filter. ACM SIGGRAPH Tutorial, 2001.
http://www.cs.unc.edu/~welch/media/pdf/kalman_intro.pdf.

Index

Printing: Krips bv, Meppel
Binding: Stürtz, Würzburg

MIX
Papier aus verantwortungsvollen Quellen
Paper from responsible sources
FSC® C105338

If you have any concerns about our products,
you can contact us on
ProductSafety@springernature.com

In case Publisher is established outside the EU,
the EU authorized representative is:
Springer Nature Customer Service Center GmbH
Europaplatz 3, 69115 Heidelberg, Germany

Printed by Libri Plureos GmbH
in Hamburg, Germany